·BASIC·
AEROBATICS

Geza Szurovy   Mike Goulian

TAB Books

Imprint of McGraw-Hill

New York San Francisco Washington, D.C. Auckland Bogotá
Caracas Lisbon London Madrid Mexico City Milan
Montreal New Delhi San Juan Singapore
Sydney Tokyo Toronto

Disclaimer

All material in this book should be used as a source of general information only. It is the responsibility of every pilot intending to learn aerobatics to receive appropriate comprehensive dual aerobatic instruction from a qualified aerobatic instructor and comply with all regulations and procedures in effect. It is the responsibility of the pilot in command to consult all official sources of information relevant to every aspect of a proposed flight and personally assure compliance with all laws, regulations, and procedures.

© 1994 by **Geza Szurovy** and **Mike Goulian**.
Published by TAB Books.
TAB Books is a division of McGraw-Hill, Inc.

pbk   4 5 6 7 8 9 10 11 12 13   FGR/FGR   9 9 8 7 6
hc   1 2 3 4 5 6   7   8   9 10   FGR/FGR   9 9 8 7 6 5 4

Library of Congress Cataloging-in-Publication Data

Szurovy, Geza.
    Basic aerobatics / by Geza Szurovy and Mike Goulian.
       p.   cm.
    Includes index.
    ISBN 0-07-062931-5 (h)   ISBN 0-07-062926-9 (pbk.)
    1. Stunt flying.   I. Goulian, Mike.   II. Title.
TL711.S8S98   1994
797.5'4—dc20   93-38046
                                        CIP

Acquisitions Editor: Jeff Worsinger
Editorial team: Bob Ostrander, Executive Editor
               Norval G. Kennedy, Editor
Production team: Katherine Brown, Director
               Rose McFarland, Layout
               Susan E. Hansford, Typesetting
               Linda L. King, Proofreading
               Jodi L. Tyler, Indexer
Design team: Jaclyn J. Boone, Designer
            Brian Allison, Associate Designer
Unless otherwise mentioned, all photographs and illustrations
  are by Geza Szurovy.
Cover: Mike Goulian flying Staudacher 300 GS   PFS
Photography by Geza Szurovy.   4445

For Anya and Apa,
who understood my love of flight, always

GS

For my brother, Matt,
the other half of the team

MG

Contents

Foreword

When people ask me how to learn aerobatics I tell them to find a good instructor and a good aerobatic book. Flying aerobatics does not require a special license or rating—with good reason. People have such different motivations for learning aerobatics, perform aerobatics as so many different levels, and have access to such varying types of airplanes that it would be hard to standardize aerobatic training for a specific flight test. A comprehensive book, on the other hand, that allows you to customize your training to meet your specific needs is indispensable.

Why do aerobatics?

Are you interested in looping and rolling in an open-cockpit biplane over green fields on a summer day? Are you interested in professional airshow flying, entertaining hundreds or perhaps thousands of spectators? How about competition flying? Are you yearning to join one of the many IAC chapters around the world and fly a contest in your Clipped-Wing Cub?

Maybe you are an airshow spectator who loves to watch your favorite performer dance across the sky in a one-of-a-kind airplane and you would like to learn more about how it is done. Or perhaps you are an R/C modeler keen to improve your flying skills.

Whatever the motivation or personal goal, aerobatic lessons will give you increased confidence in the airplane you fly. An important facet of aviation since the earliest days of flying, the precision maneuvers of aerobatics are fun to learn, exciting to perform, and guaranteed to give the pilot a challenging reward. Whether you fly a Cub or a 747, aerobatic training will increase your competence and ability as a pilot as you begin to gain a new understanding of holistic flight—flight on all axes.

Basic Aerobatics is the definitive book on modern aerobatic flying for the 1990s. It will take you from the beginning of aerobatic history to the higher levels of modern-day aerobatic competition and airshows. You will learn to appreciate how the sport of aerobatics enhances the expression of the freedom of flight and how learning aerobatics promotes discipline in your flying, and teaches you judgment within the boundaries of a safe and controlled environment. You will gain a deeper understanding of how an airplane flies in all axes and of the real—not self-imposed—limits of the equipment you fly.

Now, as airshows, aerobatic competition flying, and recreational aerobatics are ever increasing in popularity, you can take advantage, through *Basic Aerobatics*, of the decades of experience gained by aerobatic pilots before you.

Mike Goulian and Geza Szurovy are an ideal team to write this book. Mike is one of America's premiere unlimited aerobatic competition and airshow pilots; Geza is an award-winning aviation writer and recreational aerobatic pilot. Both learned to fly before they knew how to drive a car.

To write this book, Mike, who also runs his own aerobatic flight school, assumed the role of instructor and Geza acted as novice student pilot. They flew each maneuver again and again, carefully preparing for every flight in comprehensive preflight briefings and picking apart and documenting each sortie during lengthy post-flight sessions. To bring you the whole aerobatic experience, they started out in a standard basic trainer, worked up to the immortal Pitts Special, and finished up in a state-of-the-art unlimited two-seat monoplane.

As you sit down and begin your journey through *Basic Aerobatics*—a journey that will take you from the beginning of aerobatic history to today's high tech monoplanes—allow the pages to help you fly a little farther and spread your wings a little wider.

PATTY WAGSTAFF
Unlimited U.S. National Aerobatics Champion
1991, 1992, 1993

The unlimited U.S. national aerobatics championship is the highest award in U.S. aerobatics. Men and women compete for the single award; Ms. Wagstaff was the first woman to win the U.S. championship. To honor the accomplishment, the Smithsonian Institution acquired the Extra 260 monoplane that Ms. Wagstaff flew at the 1991 contest for permanent display in the National Air & Space Museum.

Acknowledgments

Mention aerobatics and the word "rugged individualist" more often than not sneaks into mind; yet aerobatics could not be what it is today without the solid cooperative spirit that is also so characteristic of the aerobatic community. Everyone we turned to was supportive, encouraging, and ready to help in every way possible. Several people went far beyond anything we dared to reasonably expect. To all we are greatly indebted.

A special thanks to Ian Groom for arranging the in-cockpit Sukhoi 29 photographs that illustrate many of the chapters, and for so patiently flying the maneuvers and accommodating the photographer's many requests; to Patty and Bob Wagstaff, Clint McHenry, and Dennis Sawyer for reading the initial manuscript and greatly improving on it with their thorough comments and observations; to Jud Milgram for so going out of his way to set us straight on matters theoretical; to Brian Becker and Sue Besarick of Pompano Air Center for their warm hospitality, sound advice, and for making available many of the best photographs; to Mike Heuer for his comments on parts of the manuscript and for directing us to the right sources of information; to Lt. Col. Marc Nathanson (USAF, Ret.) for flying some of the photo missions and for dispelling the aerobatic mysteries of air combat; to Dorothy Cochrane of the National Air and Space Museum for her encouragement and advice; to Martin Berinstein for so generously letting us take over his delightful photo studio; to Bob Schuette and the staff at Impress Design for producing the airplane silhouettes used in the illustrations; and to all the staff, past and present, at Executive Flyers Aviation.

We would also both like to thank our families for being so supportive in so many ways of our flying antics over the years.

Thanks, also, to every aviator who ever taught us anything about this best way to fly.

GEZA SZUROVY & MIKE GOULIAN
Hanscom Field, Massachusetts
May, 1993

How to use this book

In addition to helping all readers understand how the sport of aerobatics developed and what it is all about, this book is also a primer for the student of basic aerobatics. It is intended to be used as a supporting text in a basic aerobatic course given by a qualified aerobatic instructor. The maneuver chapters describe in detail the aerodynamic forces acting on the aircraft in each maneuver, how to fly each maneuver, what the common errors are and how to correct them, and what to do before safety is compromised if things go wrong.

Take the theory—the "why"—to heart. If you intimately understand what is happening to your aircraft aerodynamically at all times, aircraft control—the "how"—will quickly become second nature.

The maneuvers are presented in a sequence designed to introduce the student to progressively greater challenges in an orderly building-block fashion. Stalls are explored first, in greater detail than in nonaerobatic training, to get the student comfortable with what happens when the critical angle of attack is exceeded, and what it takes to exceed it under a variety of flight conditions. Special emphasis is placed on accelerated stalls, and stalls at various bank angles.

Rolls are introduced next, to expose the student to unusual attitudes in a low-G environment. The easiest and one of the most pleasant aerobatic maneuvers, the aileron roll, is followed by the more challenging slow roll. (The barrel roll is not addressed because it is no longer widely practiced and is not a competition maneuver.)

When the student is comfortable with rolls, it is a good time to introduce inverted flight, entered by rolling inverted. At the completion of this stage, the student will feel at home in all attitudes, will have gained experience in the delicate handling of the controls, and will be ready to move on to higher G maneuvers.

The loop is introduced next, followed by half loops and Immelmanns, which begin to combine elements of the loop and the roll. Cuban eights and reverse Cuban eights, the most complex combination maneuvers of a basic aerobatic curriculum, complete this stage of the course.

In the final stage, hammerheads (stall turns) and spins are taught. The course culminates in the student composing a basic aerobatic sequence and learning to fly it to the International Aerobatic Club's Sportsman Category competition standards in the aerobatic box.

A final maneuvers chapter briefly describes the advanced maneuvers that are the next step in the sport of aerobatics and the subject of an advanced aerobatic course to be covered in a subsequent book.

Two of the four appendices provide detailed sample curriculums of two basic aerobatics courses.

Introduction

August 19, 1913, is most often remembered by aviators because on that date the young Frenchman Adolphe Pegoud became the first person to parachute safely to earth from an airplane. He had deliberately abandoned his Bleriot monoplane at 650 feet in a daring experiment to find out if the fancy new linen escape device, successfully tested from tethered balloons, would be practical for the pilot of a heavier-than-air machine. Fortunately for Pegoud, the parachute blossomed into a full canopy and he floated gently to earth.

But what happened to Pegoud's abandoned Bleriot was equally remarkable, though less remembered. Free of human control the delicate monoplane twisted, turned, rolled on its back, and tumbled upright again and again, soaring, and diving in a fantastic aerial ballet, apparently none the worse for wear, and certainly not falling apart, contrary to contemporary expectations regarding such wild maneuvers.

Until that day, flying was largely a straight and level affair and shallow turns sent most pilots' adrenaline levels and pulse rates soaring. Some daredevils had titillated crowds with steep turns and dives, but none had deliberately rolled an airplane over on its back, or pulled its nose up past the vertical. The pilotless Bleriot's display was to change all that. By the time it crash-landed into the genteel French countryside, Pegoud was determined to duplicate the maneuvers it had performed, and aerobatics was on its way.

In quick succession Pegoud and his contemporaries mastered the basic maneuvers: the loop, the roll, the hammerhead, and by 1916 even the intentional spin. The variations on these building-block maneuvers have since been countless, nurtured by decades of barnstorming, dogfighting, airshow performances, and competition aerobatics. The result is today's flawless, gut-wrenching, eye-popping, cheek-sagging, unlimited aerobatic performance that leaves the stunned spectator wondering if there is anything these awesome machines can't do, and allows the aerobatic pilot to say that as far as the airplane is concerned, there is no such thing as being "out of control."

But just what is aerobatic flight? A wide range of interpretations run from Webster's rather unhelpful "performance of stunts in an airplane" (whatever stunts are) to the Federation Aeronatique Internationale's detailed definitions of competition aerobatic maneuvers. Most regulatory bodies worldwide consider an aircraft to be doing aerobatics whenever it deviates from the bank angles, pitch changes, and acceleration rates necessary to complete normal category maneuvers as defined or inferred by the flight manual, but the regulatory bodies are usually quite vague about where exactly the threshold into aerobatic flight is crossed. Under such a definition, any abrupt and substantial bank and pitch excursion qualifies as aerobatics, however ham-fisted or unintentionally uncoordinated.

Yet the word aerobatics implies a high degree of skill and full control by the pilot over the machine in all dimensions. So, perhaps a more useful definition is "an intentional departure from straight and level flight to fly one maneuver or a series of premeditated maneuvers that require extremes of bank, pitch, and acceleration." As Pegoud and his colleagues discovered so many years ago, the elements of aerobatic flight can be distilled to four basic aerobatic maneuvers: the roll, the loop, the hammerhead, and the spin. All other maneuvers are variations of these fundamentals.

The understanding of the fundamental maneuvers and the development of derivatives gave rise over the years to the sport of competitive aerobatics, the performance of aerobatic maneuvers to predetermined standards. The big contribution of competitive aerobatics to aerobatics as a whole is this set of standards, the benchmarks of perfection to which all properly instructed aerobatic pilots are trained and by which they are thereafter measured and measure themselves.

This book focuses on understanding and performing aerobatic maneuvers as defined by the standards of competitive aerobatics. These standards require the highest degree of precision and competence and are the foundation of basic aerobatic courses throughout the country. There is no substitute for sound aerobatic training to make a pilot understand the limits of an aircraft and teach a pilot to fly with absolute confidence throughout the performance envelope.

Many pilots who take an aerobatics course will never fly in aerobatic competitions. They might just want to enjoy romping about the skies for the fun of it on sunny summer days. They might opt for the exciting world of airshow display flying. Or they might find themselves in a jet fighter doing such combat maneuvers as the yo-yo or a high-G roll. But if they learned the basics as defined by the standards of competition aerobatics they will easily, enjoyably, and safely meet the challenges of their choice.

It might seem ironic that a sport that promises boundless freedom in all dimensions, in fact demands ironclad discipline, intense concentration, and a dogged adherence to prescribed practices and procedures. But this irony doesn't stand the test of experience and in the end aerobatics delivers its promise. Yes, the demands are ever present, but through hard work and a lot of practice, meeting them eventually becomes second nature and the airplane does indeed become an extension of your body throughout every maneuver.

In this respect aerobatics is just like any other aspect of learning to fly. Remember how little time you had to savor the freedom of flight on your first few cross-countries or solo touch-and-go patterns? You wondered what you were doing there, but you stuck with it and it paid off—and so will aerobatics. Meet its demands long enough and one day, all alone, you will do a perfect Immelmann on a whim only to realize that you did everything right without one conscious thought about it. At that moment you will truly know what aerobatics is all about.

1
Getting started

IF YOU ARE READY FOR THE CHALLENGE AND REWARDS OF AEROBATICS, IT is almost time to head for the airport and the aerobatic training machine awaiting you. But before you go, you must consider regulations, safety, the importance of getting professional instruction, selecting a training aircraft, and insurance questions.

AEROBATICS AND THE REGULATIONS

There is no escaping Federal Aviation Regulations (FARs), and it is best to clearly understand their applicability right up front. Aerobatic flight shares the airspace with all other forms of flight, so all relevant FARs covering normal nonaerobatic flight apply. In addition, there are regulations created specifically for aerobatics, and regulations that cover other special areas of aviation but also apply to aerobatics due to the characteristics of aerobatic flight. The FARs applicable to aerobatics can be divided into two general categories:

- aircraft operations
- construction and certification of aerobatic aircraft

First let's consider what the regulations do not require. There is no requirement in the United States for an aerobatic rating. It is the pilot's responsibility to seek competent aerobatic instruction prior to solo aerobatics. Accident statistics indicate that this policy is by and large effective—let's keep it that way. There is no airspace set aside by regulation specifically for aerobatic flight; as long as you stay out of airspace where aerobatics is not authorized, all you need to do is keep a good lookout and you are in business. There is no special additional medical certificate requirement; the standard medical requirements apply to the pilot in command, based upon whether the flight is private or commercial.

Operation of aircraft in aerobatic flight

Only two paragraphs of the entire Federal Aviation Regulations cover the aerobatic operation of aircraft. Only one paragraph directly and exclusively addresses aerobatic operations. It is under Part 91, Subpart D, Special Flight Operations and it is quite straightforward:

§91.303 Aerobatic Flight

No person may operate an aircraft in aerobatic flight-

(a) Over any congested area of a city, town, or settlement;

(b) Over an open air assembly of persons;

(c) Within the lateral boundaries of the surface areas of Class B, Class C, Class D, and Class E airspace designated for an airport;

(d) Within four nautical miles of the centerline of any federal airway;

(e) Below an altitude of 1,500 feet above the surface; or

(f) When flight visibility is less than three statute miles.

For the purpose of this section, aerobatic flight means an intentional maneuver involving an abrupt change in an aircraft's attitude, an abnormal attitude, or abnormal acceleration, not necessary for normal flight.

Fig. 1-1. *The classic Stearman evokes an age that is long gone.*

Patty Wagstaff Airshows

Fig. 1-2. *Aerobatics unlimited: U.S. National Champion Patty Wagstaff putting her Extra 300S through its paces.*

The other aerobatic paragraph applies to the use of parachutes (§91.307). It was not written exclusively for aerobatics, but is equally important because it specifically defines the conditions under which parachutes must be carried and the intervals and conditions under which they must be periodically repacked. For aerobatic flight the bottom line is that if there is more than one person in the aircraft and a bank angle will exceed 60° or a pitch angle will exceed 30° relative to the horizon, parachutes must be worn by both occupants. Specifically, §91.307(c) reads:

Unless each occupant of the aircraft is wearing an approved parachute, no pilot of a civil aircraft carrying any person (other than a crew member) may execute any intentional maneuver that exceeds-

    (1) A bank of 60 degrees relative to the horizon, or

    (2) A nose-up or nose-down attitude of 30 degrees relative to the horizon.

3

Fig. 1-3. *Former European championship competitor Ian Groom practices his airshow routine in his brand-new Sukhoi SU-26M.*

The next subparagraph exempts flight tests being given for a rating, instruction in spins, and other maneuvers required by the regulations given by an appropriately authorized instructor. There is also an exemption if the second person is a legally required crew member, but in most cases this does not apply to civilian aircraft licensed for aerobatics because practically none of them have a legal requirement for a second crew member. The parachute repacking requirement of §91.307 is self-explanatory.

Other regulations apply as they do to normal, nonaerobatic flight. Because many aerobatic aircraft are taildraggers, it is especially worthwhile to review taildragger checkout and currency requirements.

Upon reading the regulations, aspiring aerobatic and airshow display pilots might wonder how aerobatics can be flown at competitions and airshows, where performances are routinely flown below 1,500 feet agl and in the vicinity of large crowds, and how competition aerobatics can be practiced at the required low altitudes. Such activity is made possible by waivers obtained from the FAA specifically for each event and by the establishment of FAA-approved aerobatic practice zones. A highly detailed proposal must be submitted describing the practice zone or proposed event. For proposed events, the information must include details on flight paths, crowd areas, participating acts, and flight crew experience. The FAA also conducts site visits and if a waiver is granted for an event, inspectors will be on hand to see that everything goes as agreed.

It is interesting to note that strictly for the purpose of considering an aviation event waiver, the FAA's *Inspector's Handbook of Guidelines* instructs the users to take a narrower interpretation of aerobatics than is provided in the FARs. Inspectors are advised that for an airshow, aerobatics is inverted flight and all the standard aerobatic maneuvers, such as slow rolls, snap rolls, loops, Immelmanns, Cuban eights, hammerhead turns, and the like, which cannot be performed over congested areas or spectators.

Fig. 1-4. *The delightful little Decathlon, first aerobatic mount of many pilots.*

Certification of aircraft for aerobatics

The regulatory definitions of aerobatics and the conditions under which you may fly aerobatics in the airspace is only one aspect of the FARs in which you must be well versed. It is equally important to know what aircraft may be legally flown aerobatically.

Production aircraft must be approved for aerobatics during the certification process. The requirements the aircraft must meet are covered in FARs 21 and 23. In addition to being approved in a specific aerobatic category, the aircraft must also be approved for each specific aerobatic maneuver it is to fly. For general instructional purposes, some aircraft not approved in the aerobatic category might be approved for a limited number of maneuvers that technically qualify as aerobatics, such as spins; thus, in any production airplane, you may fly only those aerobatic maneuvers that are specifically authorized in the aircraft operating manual.

For an insight into the differences between aircraft licensed in the normal category and the aerobatic category, consider the Cessna 150 and the Cessna Aerobat. Superficially the aircraft look alike, and the Aerobat was derived from the 150. Yet their operating manuals clearly reveal that from an aerobatic standpoint they are completely different aircraft. The Aerobat is designed to withstand considerably greater loads than the 150, allowing it to safely fly maneuvers that could easily overstress the 150.

Approved aerobatic maneuvers

1975 Cessna 150
Flight load limits:
(Flaps up) +4.4G & −1.76G; (flaps down) +3.5G
Chandelles
Lazy eights (Continued on page 6.)

(Continued from page 5.)
Steep turns
Spins
Stalls

1977 Cessna Aerobat
Flight load limits:
(Flaps up) +6.0G & −3.0G; (flaps down) +3.5G

Chandelles
Lazy eights
Steep turns
Spins
Stalls
Loops
Cuban eights
Immelmanns
Aileron rolls
Barrel rolls
Snap rolls
Vertical reversements

Only an unthinking pilot would try anything but chandelles, lazy eights, stalls, and spins in a 150, and he would be breaking regulations.

While the question of doing aerobatics in a certified production aircraft is quite straightforward, in the case of experimental aircraft, more latitude is left to the designer and builder. For experimental aircraft, no regulatory requirements specifically authorize aerobatic maneuvers, let alone list them in the aircraft operating manual; however, many homebuilt aircraft are as good or better than factory models. Many homebuilt designers and builders take great pride in their work and know and document the capabilities of their aircraft in great detail. The threat of liability suits is another reason for the experimental builder to be thorough.

Let the buyer beware. Choose a tried design, and if you are buying a finished experimental aircraft rather than building your own, carefully check out the builder and the airplane's history, including its aerobatic activities.

SAFETY AND THE IMPORTANCE OF PROFESSIONAL INSTRUCTION

In aerobatic flight, you deliberately operate your aircraft closer to the extremes of its performance envelope, the limits of airspeed and load, than during any other form of flying. The closer you are to these limits, the easier it becomes to inadvertently exceed them. As you hover on the edge of a stall, an indelicate or misguided control movement can push you over the edge and perhaps even into an unintentional spin. Close to redline and accelerating, an excessively urgent pull on the stick can send the G loads on your aircraft and your body soaring beyond tolerance. And misjudging the altitude required to complete a maneuver can really ruin your day. So you must take precautions and develop and adhere to sound safety habits.

The right attitude

In large measure safety is a matter of having the right attitude. All the regulations, policies and procedures, and credible advice in the world will not keep you safe if you don't intend to abide by them.

Get professional instruction

Aerobatics is a mature sport with decades of tradition borne of experience. Anything that the novice aerobatic pilot can try has been developed to perfection by others before and is safely made available to newcomers by qualified aerobatic instructors. Only an unthinking pilot would endanger body and soul and everyone else within reach by attempting self-taught aerobatics. Make it your business to learn aerobatics only from a properly qualified aerobatic instructor. Self-study is strewn with pitfalls that make it an unacceptable option.

Though they might think otherwise, the self-studying novices lack the skills to consistently stay within the aircraft's performance limits and to save themselves when they surely overstep those limits. Nor do they have any point of reference to know whether their antics have any resemblance to the precision aerobatic maneuvers they claim to be doing.

Self-study usually leads to one of two things: either the pilot gets scared sufficiently to stop further experiments, and maybe abandon aerobatics altogether, or becomes an NTSB statistic. Do yourself and the sport a big favor. Get a qualified aerobatic instructor.

Where to find aerobatic instruction

Having committed yourself to learning aerobatics from a qualified school or owner-operator, the next question is where to find one. Because there is no aerobatic rating requirement, or any special additional regulations of flight schools and instructors offering aerobatic instruction, how can you evaluate what you get? You need to do some homework, but the process is easier than it might seem.

A good place to start is the International Aerobatic Club (IAC), the Experimental Aircraft Association's aerobatic arm. The IAC maintains a list of aerobatic schools throughout the country (though it does not evaluate or endorse them), and can provide the names of schools offering aerobatic instruction in your area. The IAC's address is in appendix C. The IAC also has a large number of chapters nationwide, the grass roots support group of the sport. It is worth looking up the local chapter in your area if you plan to do aerobatics. You will find excellent advice, reliable leads regarding instruction, and great camaraderie.

Other sources of information are the advertising sections of aviation magazines and newspapers, and the local airport grapevine. Some nationally known aerobatic centers that can be easily reached through their advertising in the aviation media might also be able to get you in touch with a school in your area.

When you have located a school or owner-operator, check it out thoroughly. Ask for a description of the course, see what aircraft are used, and find out how long the establishment has been in business. Ask how many students it has, what the aerobatic experience of the instructors is, do they do any other form of aerobatic flying such as competition or display flying, and whatever else might come to your mind. Talk to some former students about their experience and if everything checks out, sign up for an introductory lesson.

Safety pointers

When you chose to seek qualified aerobatic instruction, you took an important step toward learning aerobatics safely. But as in all other forms of flying, it is important to maintain a safety conscious attitude throughout your aerobatic flying. Though by no means exhaustive, and intended to complement general safety practices, here are some safety pointers particularly apt for the aerobatic pilot.

Understand aerobatic theory. It is by understanding not only "how" and "what," but also "why" that you will find it effortless to stay ahead of your aircraft during all phases of aerobatics. During flights to performance limits, your grasp of theory and your ability to apply it might mean the difference between being safe or overstepping your airplane's capabilities.

Know your airplane. This is a basic requirement and it means knowing not only performance parameters, but all of the aircraft's behavioral characteristics, the design features responsible for them, and the designer's thinking behind them (again, think in terms of "why," not only "what").

Always have an out. Learn to think in terms of contingencies. Make it a point to study the ways out of trouble and be ready with your options if a maneuver goes wrong.

Wear a parachute. Some pilots claim that they would never jump, and under certain circumstances, such as during low-level airshow flying, the chance of a successful parachute jump might be slim. But most of your aerobatics will be at altitude, so why deny yourself an option if all it takes is strapping it on? Also, be sure to comply with the legal requirements of wearing parachutes.

Always have plenty of altitude. The FARs' minimum altitude for aerobatics is just that: minimum. Based upon aircraft performance characteristics and contingencies for botched maneuvers, an extra altitude margin is always a good idea. As they say, one of the most useless things in aviation is the altitude above you. For some types of aircraft, such as high-performance warbirds, even as much as 5,000–6,000 feet might not be enough to safely recover from certain botched maneuvers.

Fly only in good visibility. Many aerobatic pilots consider the 3-mile minimum visibility requirement as insufficient for aerobatics because it leaves too little time to notice and evade conflicting traffic while flying an aerobatic sequence. A commonly used personal standard is 7 miles visibility.

Always check maneuver entry speed and altitude. Make a habit of glancing at the airspeed indicator and altimeter prior to any aerobatic maneuver to verify that you have sufficient speed and altitude for the maneuver. This point might seem obvious, but it will prove helpful when you begin to string sequences together and will be tempted to zoom from one maneuver to the other without a quick check of airspeed and altitude.

Practice only what you know. As you make progress and gain confidence during your training, you might be tempted to try new maneuvers that you haven't yet learned on your own, which could get you in trouble. It is best to be conservative and check with your instructor before you try anything solo.

Be rested and in good health. Flying aerobatics when you are tired or ill is a sure way to lose your edge and compromise your safety.

Respect FARs. It seems fashionable to complain about regulations, but most of them make good sense, and in many respects the amount of regulations is actually quite minimal. Bear in mind that new regulations are frequently the result of someone not towing the line of unregulated common sense, prompting the Feds to regulate.

HEALTH AND PHYSICAL CONDITION

Basic aerobatics does not require the pilot to be in extraordinary physical shape. Anyone in basically good health who can pass the FAA Class III physical for the private pilot's certificate should feel quite comfortable doing basic aerobatics after a chance to get used to the new sensations; however, it is true that the better shape you are in, the less you will feel the effects of aerobatics, and to withstand the punishing maneuvers of high-G advanced aerobatics, good physical conditioning is essential.

To increase G tolerance, weightlifting is the best exercise—better than cardiovascular exercises that improve circulation, making it easier for the blood to flow through the body and collect in certain body areas to the detriment of others during high G maneuvers.

A concern of most novice aerobatic pilots is airsickness. Fortunately, in most cases nausea appears to be caused psychologically, usually from a fear of what is to come. As soon as the pilot ceases to tense up the instant that the aircraft commences an aerobatic maneuver and the pilot's body is conditioned to flow with the airplane instead of trying to fight it, nausea ceases to be a factor. Many people feel nauseous when someone else is doing the flying; as soon as they are given the controls, their condition improves. Tolerance can usually be developed over time. It also helps not to overeat before an aerobatic session. For a small minority of pilots, nausea has physical causes, in which case it is best to consult a flight surgeon.

An often overlooked, but common factor in flying aerobatics comfortably is the importance of being well rested before an aerobatic session. Doing any serious aerobatics with a hangover would be especially counterproductive, to put it mildly. For some pilots, eating habits can also have an effect on their comfort level during aerobatics. It is usually best to avoid an empty or overstuffed stomach prior to an aerobatic session. Scrub the flight if you feel the least bit unwell, such as when you have flu symptoms. Be especially wary of any ear or sinus trouble. The aerobatic pilot should understand the effective G forces on the body. This topic is addressed separately, in chapter 2, following a general discussion of G forces.

SELECTING AN AEROBATIC TRAINING AIRCRAFT

Aerobatic aircraft come in a surprising number of configurations: high-wing, low-wing, monoplanes, biplanes, tandem seating, side-by-side seating, yokes, sticks, fixed-pitch propellers, constant-speed propellers, taildraggers, tricycle gear, and on and on. What to choose is a matter of personal preference, and your shortlist may also be determined by what is available in your area. Nevertheless, a few general observations are worth thinking about.

Pure aerobatic aircraft or derivatives? It is generally preferable to choose an aircraft that was designed specifically for aerobatics, rather than one designed for other purposes and certified in a modified version for aerobatics. Often the modifications strengthen the structure to withstand higher Gs, but otherwise leave the aircraft unchanged. Many of the modified aircraft will thus have airfoils and control systems designed without aerobatics in mind, which often results in uninspiring handling characteristics barely adequate for basic maneuvers. On the other hand, an aircraft specifically designed for aerobatics is always more likely to be light and crisp on the controls in comparison. For example, aerobatics in a Decathlon are far more pleasant and productive than in a Citabria. Another advantage of aircraft specifically designed for aerobatics is that they are more likely to have inverted oil and fuel systems.

High wing, low wing, monoplane, biplane? This question is entirely a matter of personal preference. Aerodynamically, there is little difference from the standpoint of basic aerobatics. For many people, the fact that they are doing aerobatics in a biplane such as a Pitts is an added thrill, evoking an earlier era of aviation. Others feel that the extra wing blocks too much visibility. All high-wing aircraft tend to have rather restricted visibility. It is probably fair to say that when you have been in a bubble-canopied low-wing monoplane, its stunning visibility will be hard to give up.

Tandem or side-by-side? Tandem seating allows you to sit on the aircraft's longitudinal axis, enabling you to get a more precise sense of your position and trends during a maneuver. On the other hand, side-by-side seating makes communications much easier between student and instructor, especially if hand gestures are called for. For the recreational pilot, either option is probably fine, while the student intent on competition aerobatics might opt for tandem seating from the outset.

Stick or yoke? Stick!

Fixed-pitch or constant-speed? Doing aerobatics in an airplane equipped with a constant-speed propeller is great because you don't have to worry about engine overspeed, and in most maneuvers, even the throttle setting is not worrisome. To review, a constant-speed propeller adjusts its blade angle in changing flight conditions to maintain a constant RPM. In many cases, you can just set up the power at the beginning of the session and leave it untouched throughout, allowing you to fully concentrate on the maneuvers. A fixed-pitch propeller's RPM varies with changing flight conditions. With the fixed-pitch propeller, RPM increases with aircraft speed and, if not checked, might easily result in an engine overspeed. On airplanes with fixed-pitch propellers, you frequently find yourself having to make distracting power adjustments and casting worried glances at the tachometer; however, learning aerobatics with a fixed-pitch propeller is good training to sensitize you to monitoring and coping with the limits of an airplane.

Basic or high-performance trainer? It has become more and more the practice to bypass aerobatic trainers and jump directly into the dual version of a high-performance competition airplane. While this is not inherently undesirable, you will miss something if you leave the low-performance aerobatic trainer out of your curriculum. It is more difficult to fly aerobatics well in an airplane of lower performance. If at first you opt for the less-sexy trainer, you will be a better pilot. Perhaps the best approach is the progressive upgrade to machines of higher and higher performance. Say, start with a Decathlon, followed by a Pitts, and then a Staudacher, a Sukhoi, an Extra, or any other unlimited monoplane.

Regardless of your choice, after your initial training, when you go on to fly aerobatics in aircraft of either lower or higher performance, a thorough dual aerobatic checkout in each aircraft should be the first item of business; if it is a single-seater, get checked out in a two-seater of very similar flight characteristics.

AEROBATICS AND AVIATION INSURANCE

The law-abiding aerobatic pilot need not worry much about special insurance requirements for aerobatic flight because usually there are none. Insurance is generally written to cover whatever type of flight is authorized by the flight manual, and in the case of aerobatic aircraft, this includes aerobatic flight. To be fully informed of your specific situation, check your policy and consult your insurance provider.

The case of an experimental aircraft might be a bit more murky, especially if the aircraft operating manual does not specifically address aerobatics. The prudent pilot will carefully research the suitability of the experimental aircraft for aerobatics and will confirm coverage of aerobatic flight with the insurance provider.

An easy way to invalidate insurance coverage is to violate the FARs, including the terms of your airplane's airworthiness certificate. If you come to grief flying aerobatics in an aircraft not appropriately certified, your insurance coverage will most likely be invalid. An accident or incident during aerobatics below a legal minimum altitude might also compromise coverage.

It is also a good idea to make an extra effort to understand those regulations that affect aerobatic flight, but that you might not regularly encounter during your normal nonaerobatic flying. Examples are regulations covering the use and maintenance of parachutes, and the checkout and currency requirements for flying taildraggers. Bear in mind that any damage or injury caused to third parties will always open you to a hefty liability lawsuit and if your insurance coverage is inadequate or you caused the problem while busting FARs, you will place not only yourself but your family in grave financial jeopardy.

Aerobatic pilots who intend to participate in airshow displays or in competitions with or without compensation should consult their insurance provider to check what special insurance conditions might apply.

2
Loads and limits
on machine and pilot

ONE REASON FOR LEARNING AEROBATICS IS TO BE A BETTER PILOT. PART of being a better pilot is understanding not only what happens to your aircraft in aerobatic flight but also why things happen the way they do. Only by understanding the theory behind the maneuvers can you form a mental picture in flight of where your aircraft is on the performance envelope, where it is headed, and what you need to do to fly a maneuver as precisely and closely to perfection as possible and stay within performance limits.

The good news is that you have seen this material in ground school when you studied for your pilot's certificate's written exams. All you need to do is refresh your memory, take a closer look at the elements that are important to aerobatic flight, and then go up and deliberately experience in the aircraft the practical effects of what you learned. The intent is not to tax you with physics and math, but to help you grasp in layman's terms the theoretical concepts that define the aerobatic pilot's universe and provide a foundation of safe, confident aerobatics.

AEROBATICS IS ENERGY MANAGEMENT

To operate your aircraft to the limits of its performance, you have to properly manage its energy, represented by altitude and airspeed. Energy is generally distinguished as *potential energy* and *kinetic energy*. You can think of potential energy as energy represented by altitude, available for conversion into airspeed. In the same vein, you can think of kinetic energy as energy represented by airspeed, available for conversion into altitude. During a typical aerobatic maneuver, an increase in kinetic (airspeed) energy is accompanied by a decrease in potential (altitude) energy, and vice versa.

To illustrate this point, consider the inside loop (FIG. 2-1). At the beginning (bottom) of the loop, airspeed (kinetic energy) is high. By pulling up into the loop, the pilot converts kinetic (airspeed) energy into potential (altitude) energy. As the aircraft "floats" over the top of the loop, kinetic (airspeed) energy is at its minimum and potential (altitude) energy is at its maximum. The *total energy*, (the sum of kinetic and potential energy) is essentially the same as it was at the beginning of the maneuver (except for the minimal effect of drag, which can be ignored for these purposes). As the aircraft completes the second half of the loop, airspeed increases to its original value (actually a few knots less due to the effect of drag) and altitude decreases to its original value. The potential energy gained upon reaching the top of the loop has been converted back into kinetic energy. In an outside loop flown from upright straight and level, the same relationship exists, except the aircraft starts and ends the maneuver at the point in the cycle where its potential (altitude) energy is at its highest value.

To complete an aerobatic maneuver not only must you have to have sufficient kinetic energy to begin with (or potential energy convertible into kinetic energy), but you also have to manage it in a way to leave you sufficient kinetic energy for all segments of the maneuver. If, for example, your pull-up into a loop is so gentle and wide that by the time your nose goes through vertical your airspeed hovers near stall speed, you will not have sufficient kinetic energy to complete the maneuver. At some point, your airspeed will drop off to zero, and the effects of slipstream, torque, and gyroscopic precession will make the airplane do interesting things on its own to get itself out of the mess you got it in, unless you have had the proper instruction to salvage the situation. You will have mismanaged the aircraft's energy by failing to convert kinetic (airspeed) energy into potential (altitude) energy rapidly enough to complete the maneuver. The same thing will happen if you use up too much airspeed taking your time to claw your way up the second half of an outside loop.

If you squander kinetic (airspeed) energy on the way up in any maneuver, in order to gain it back, you will have to dive to a point below where you started, resulting in a net altitude loss. In terms of converting potential energy into kinetic energy, the aircraft has to dive the same distance, but it will start from a lower altitude relative to the ground; therefore, it needs to get below the starting altitude, which is sloppy for the recreational pilot, and a big fat zero in competition.

These examples reveal another important fact: Not only is aerobatics energy management, but the airspeed indicator serves the pilot as a measure of the kinetic energy available at any point in the maneuver. Every maneuver in every aerobatic aircraft has a minimum airspeed value representing the minimum amount of kinetic energy required to complete it with little or no net altitude loss (refer to the aircraft operating manual).

Fair enough. Then why don't we build up a big head of steam, roar way beyond the speed required for a particular maneuver, and never run out of the kinetic energy required? This practice is undesirable because aside from making the control forces on the stick uncomfortably high, minor changes in control inputs or the impact of gusts on the aircraft could result in loads that would overstress the airframe. This point brings

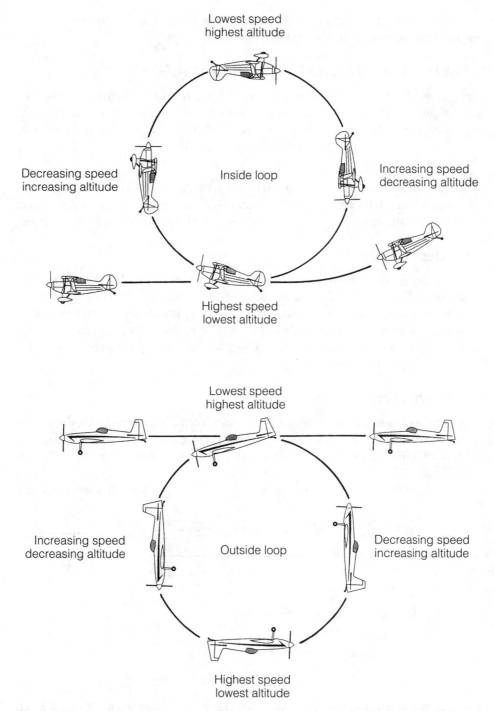

Fig. 2-1. *Aerobatics is energy management—the tradeoff between airspeed and altitude.*

us to another critically important topic in aerobatics, the airframe's structural load limits. But first we must discuss some basic physical concepts.

SPEED, VELOCITY, AND ACCELERATION

Confusing talk of "Gs" abounds when hangar flying turns to aerobatics, yet the concept is quite straightforward. Let's review some basic concepts from mechanics, the part of physics that deals with the motion of objects when subjected to various forces.

Speed

Speed is how fast you are going, measured relative to a fixed observer (for example, your favorite aerobatics judge). The direction doesn't matter. An airplane flying at 100 knots has the same speed whether it's flying 100 knots north, south, up, or down.

Velocity

Velocity is how fast you are going, *and* in which direction. Velocity is different from speed. For example, if you made a hard turn in your car without slowing down, your speed would remain unchanged, but your velocity would change. Things, such as velocity, that have both a magnitude and a direction are called *vectors*. Engineers like to think of vectors as arrows of varying length (the length representing speed, the placement of the arrow, direction).

Acceleration

Acceleration is the rate at which the velocity vector is changing. Acceleration is also a vector quantity, as can be seen by considering the case of an airplane starting its takeoff roll (FIG. 2-2). At a given point in time, the airplane is traveling down the run-

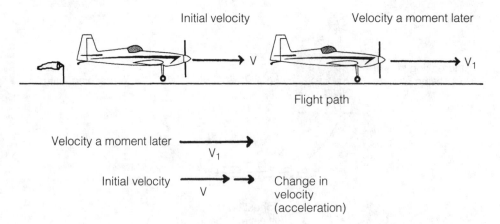

Fig. 2-2. *Linear acceleration: The velocity change along a flight path. Direction is unchanged, airspeed changes.*

way at a given speed and its velocity can be represented by the arrow shown in the figure, pointing straight down the runway. A short moment later (say, 1 second) the airplane will be traveling slightly faster, and the arrow representing its velocity will grow slightly longer. By comparing the two vectors, we can see that the change in the vector is also a vector pointing in the direction the aircraft is traveling. If we divide this small change in the velocity vector by the short time interval, we get the acceleration.

Linear acceleration. In the example above, the acceleration lies in the direction the object is traveling. Such acceleration is known as *linear acceleration*. We experience this kind of acceleration when we start our takeoff roll, brake during rollout, ride in building elevators, and the like. Here the important fact is that the magnitude of the velocity (the speed) is changing; the direction remains the same.

Normal acceleration. Acceleration also occurs when the speed remains the same, and the direction of the velocity changes. Consider the airplane pulling out of a dive in FIG. 2-3. If we compare velocity vectors from one moment to the next, we find that

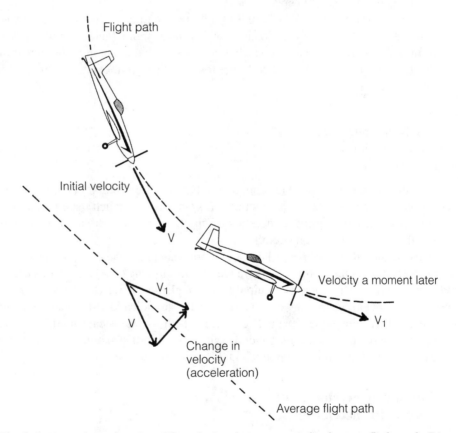

Fig. 2-3. *Normal acceleration: The velocity change perpendicular to a flight path. Direction changes, airspeed is unchanged.*

their magnitudes (speeds) do not change (in this case at any rate), but their directions do. What is more interesting is that the change in the velocity vector does not lie along the flight path, but perpendicular to it; thus, the aircraft at the bottom of a pull-up is accelerating upwards. Similarly, a car or an airplane in a steady turn is accelerating toward the center of the circle in which it is traveling. This sort of acceleration lying at right angles to the path of travel is called *normal acceleration*. (There's nothing abnormal about the other kind; "normal," meaning "at right angles," comes from the Latin word for "carpenter's square.")

Normal acceleration is the force we feel when we refer to "pulling Gs." It is proportional to the square of the airspeed and the inverse of the radius of curvature (the "tightness") of the pullout. The bottom line: the higher the airspeed and the tighter the pullout (or pushover) the higher the normal acceleration.

Turns are essentially the same as pull-ups in this regard. A higher speed turn or a tighter turn radius will require a higher acceleration toward the center of the turn. An F-16 at 400 knots maintaining 3Gs in a turn will be flying a much wider turn than a Pitts maintaining 3Gs at 120 knots.

Newton's second law. We take the trouble to carefully define acceleration because it is a direct function of the loads to which the airframe (and our body) is subjected in flight. A long time ago it was observed (by Newton, among others) that the acceleration of a body can be related to the force applied to it, expressed by the following formula:

$$\mathbf{F} = m \times \mathbf{a}$$

where $\mathbf{F}$ is an applied force,
$\mathbf{a}$ is the acceleration, and
m is the mass of the body

$\mathbf{F}$ and $\mathbf{a}$ are written in boldface to remind you that they are vectors. Force is a vector because it has a magnitude and a direction. *Mass* means how much material the body contains (for practical purposes, the mass of an object is proportional to its weight, such as the gross weight of an aircraft).

This law says that the force and the acceleration are in the same direction, and they are proportional to one another. So, if we require the airplane to accelerate in a certain manner, we know what force is required to accomplish it. If we are in a pull-up and know the normal acceleration (see below), we have an indication of the aerodynamic force normal (at right angles) to the flight path—the lift—that is causing this acceleration. Obviously, there is a limit beyond which the aircraft structure is incapable of withstanding the load imposed upon it and the structure breaks.

Measuring acceleration:
Gs and the accelerometer

Acceleration is measured in multiples of the acceleration of gravity (G), which is the acceleration of an object falling toward the earth's surface, about 32 feet per second per second. "1G " means that the lift on the aircraft is exactly equal to the aircraft's weight (the force with which the earth's gravity attracts the aircraft's mass). The *accelerometer* (or *g-meter*) is an instrument used to measure acceleration, and thus gives the pilot a rough indication of the wing lift and resulting structural loads being imposed upon the airframe. In level flight it reads 1G, indicating that the aircraft's lift is equal to its weight. In a 6-G pull-up, the 6G indicates that the lift on the wing is six times the aircraft's weight. The multiple of G is also called the *load factor* (6G = load factor of 6).

The accelerometer's most obvious use is to ensure that you do not overstress your aircraft or exceed its limit load factor, which is further explained in the flight envelope subsection of this chapter. Most accelerometers have pointers that indicate the maximum positive and negative load factors experienced in a particular flight.

The accelerometer also tells you if your acceleration matches the acceleration required to perform the maneuver that you are flying. As you gain more experience you will learn to judge Gs by the seat of your pants, but it is good to occasionally glance at the accelerometer for confirmation. Before we examine the limits of the loads sustainable by the aircraft structure, let's take a quick look at handling acceleration from the cockpit.

Controlling acceleration

To better grasp how to handle acceleration in flight, simply think of what happens in the cockpit. You feel acceleration only as long as you initiate a pitch change, that is, only as long as you are pulling or pushing on the stick. The harder you pull or push the more acceleration you will induce. But as soon as you release the stick back to neutral, you cease acceleration, and begin to reestablish equilibrium. Acceleration will immediately lighten, and eventually return to 1G .

By the same token, if you roll into a tight turn requiring back pressure on the stick, the acceleration you are creating will push you into your seat. Return the stick to neutral, and the G forces diminish. The same thing is true in reverse for outside maneuvers (the more quickly and farther you push the stick, the higher the negative Gs). From the effects of these stick movements comes the expression of "loading" and "unloading" the airframe (with acceleration expressed in Gs), and an awareness that to unload an aircraft, just move the stick toward neutral.

STRUCTURAL DESIGN REQUIREMENTS

FAR Part 61 prescribes certain requirements a pilot must fulfill before being granted a pilot certificate or rating. Similarly, other regulations prescribe the loads to which an aircraft must be designed before it can be granted a standard airworthiness certificate. Light aircraft are covered in FAR Part 23. To better understand structural limits, let's look at the structural design requirements in FAR 23 that concern the aerobatic pilot. If you want to design an aircraft, that task is beyond this book; therefore, refer directly to the regulations—and good luck!

Parke's dive

When we think of progress in aircraft control, we usually conjure up images of steely eyed test pilots deliberately putting an airplane through its paces to discover what it will do and how to handle it. Yet more often than we might think, the discovery of a certain maneuver might be quite accidental. Such was the case with what came to be known during the pioneering days of aviation as *Parke's dive*.

In 1912, Lieutenant Wilfred Parke was completing a series of routine maneuvers near Larkhill Aerodrome in England in an Avro G cabin biplane and was setting up a spiraling approach back to the airfield. He must have gotten a little slow and a little nose high because at about 800 feet agl, the left wing suddenly fell out of the sky, the nose dropped alarmingly, and Lieutenant Parke was staring straight at some rapidly rotating grass. He had inadvertently fallen into a spin.

Terrified as he was, Parke had the presence of mind to instantly attempt recovery by systematically applying a series of control maneuvers that he thought might save him. As he recounted in *Flight* magazine, he first applied full throttle in the hope that the propeller would pull the nose up. When it didn't, he retarded the throttle, applied full left rudder and pulled the stick full aft in the hope of initiating a tight left turn to recover—all the wrong moves and what an insight into the astonishingly low level understanding at that time regarding aerodynamics.

Parke was rapidly running out of ideas and altitude at the same time. To brace himself against the imminent crash, he let go of the stick, which instantly moved forward into neutral on its own accord, and in a last desperate effort to counter the leftward rotation of the aircraft he applied full right rudder at an altitude of 50 feet. The airplane immediately recovered. Parke reentered the pattern and made an uneventful landing.

In spite of himself, and without understanding the underlying aerodynamic reasons, he had stumbled into the standard spin recovery technique: opposite rudder to stop the rotation and forward stick to decrease the angle of attack and unstall the inside wing.

Parke's recovery technique was duly noted and widely publicized. Its application resulted in an increasing number of successful recoveries from inadvertent spins. But the ignorance of the underlying aerodynamics and the consequent fear of the maneuver was so high that scientific spin research through the deliberate and repeated spinning of aircraft was not undertaken until 1916.

Keep in mind that aircraft with "Experimental" airworthiness certificates do not necessarily conform to the requirements of FAR 23. If you fly such an airplane, you must rely on information provided by the designer. Also bear in mind that some foreign aircraft might be subject to slightly different design requirements than U.S. aircraft.

Flight envelope

The famous *envelope*, also known as the V_n *diagram*, is a graphic depiction of the load factor and velocity limits for a given aircraft. The shape of the graph somewhat resembles an envelope. The regulations define a flight envelope for symmetric maneuvers (wings level pull-ups, pushovers, and the like). The aircraft must be strong enough to fly every point in this envelope. A simplified envelope is shown in FIG. 2-4. The envelope is actually a combination of two envelopes; the *maneuvering* envelope is depicted together with the *gust* envelope. In cases where the area of the gust envelope exceeds the area of the maneuvering envelope, the excess area is clearly identified.

The *maneuvering envelope* is quite simple; it arises from the need to maneuver the aircraft in symmetric pull-ups and pushovers to certain load factors. For aerobatic aircraft, the regulations require these *limit maneuvering load factors* to be at least +6G to −3G . Many aerobatic aircraft are designed to even higher loads.

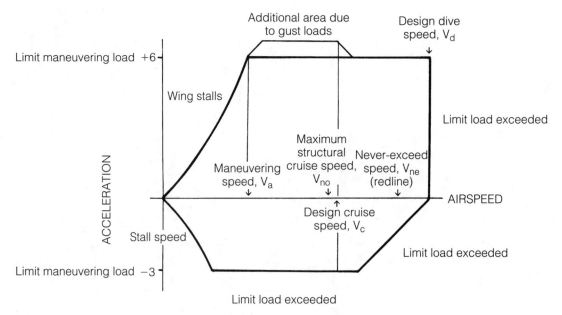

Fig. 2-4. *The envelope.*

The maneuvering envelope is bounded on the top and the bottom by the limit maneuvering load factors. The left side of the envelope is limited by the airplane's *stall speed*, V_s. This is the wings-level stall speed with the flaps (if any) retracted. The upper left and lower left corners of the maneuvering envelope are cut to reflect the fact that at a given airspeed the wing can produce only so much lift before it stalls.

The *maneuvering speed* (V_a), is defined in the regulations by the following formula:

$$V_a = \text{stalling speed} \times \sqrt{\text{limit positive maneuvering load factor}}$$

This is a simple formula for the speed at which the wing is capable of producing the limit maneuvering load factor without stalling.

What does maneuvering speed mean to the aerobatics pilot? It is a speed at which the designer must design for loads due to control inputs. Specifically, at V_a:

- The vertical stabilizer and the entire airplane must be designed to the loads resulting from a sudden maximum rudder input.

- The ailerons and wing must be designed to the loads resulting from a sudden maximum aileron input.

- The horizontal stabilizer must be designed to the loads resulting from a sudden maximum forward or aft elevator input.

The last item deserves further discussion. Note, there is no specific requirement that the wing be designed for the loads resulting from a full and abrupt elevator input at V_a. True, the formula for maneuvering speed would seem to suggest that below V_a the wing will stall before the limit load is reached, but this is not necessarily the case. For one thing, the formula applies to aft elevator inputs and positive load factors—in many aircraft the negative limit load factor is lower than the positive limit load factor and could be exceeded in an aggressive pushover at airspeeds below V_a. Second, the regulations allow the maneuvering speed to be based upon a calculated stall speed. The actual stall speed might end up being slower than this. Finally, the formula gives a *minimum* value of V_a. The regulations allow designers to set V_a higher than this value if they choose. They might indeed have their own reasons for doing so.

The point of all this is that you should not consider maneuvering speed an invitation to fly at V_a and abruptly pull (or push) the stick to its stop. You may apply full and abrupt control inputs, but it is still up to you to make sure you don't exceed the limit load factor. For many aircraft, different maneuvering speeds are listed for different gross weights. At a lighter weight, it takes less wing lift to produce the limit load factor (remember $\mathbf{F} = m\mathbf{a}$; therefore, $\mathbf{a} = \mathbf{F} \div m$). For example, to get a constant acceleration ($\mathbf{a}$) of 6, if you reduce (m) you also need less $\mathbf{F}$ (lift): $12 \div 2$ and $36 \div 6$ both equal 6 (FIGS. 2-5 and 2-6). At a lower weight, a slower speed will generate the lower lift to yield the same acceleration; therefore, the maneuvering speed must be reduced. In the formula for V_a, the stall speed varies with the weight in just the right way to compensate for this.

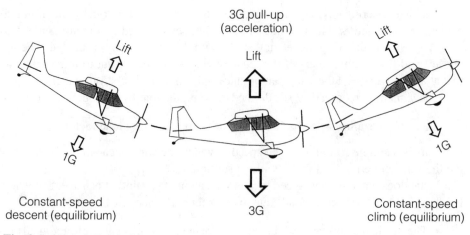

Fig. 2-5. *Upright 3G pull-up.*

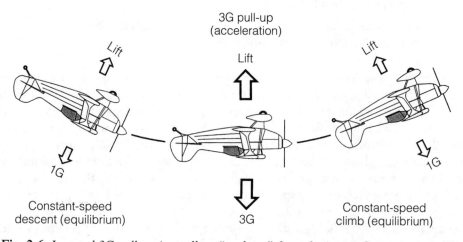

Fig. 2-6. *Inverted 3G pull-up (actually a "push-up" from the inverted perspective). Note that acceleration (G) acts in the same direction as in the upright pull-up in the preceding illustration; however, because the aircraft is upside down, on the aircraft structure acceleration acts in the opposite direction, the aircraft is said to experience negative G.*

There are maneuvers with entry speeds faster than maneuvering speed. This should be of no great concern as long as you observe the limit load factors, avoid handling the controls abruptly and use less than maximum control deflections. It is difficult to say just how much control deflection is allowed, but in the case of symmetric maneuvers (pull-ups, pushovers) you should be safe if you remain within the flight envelope. Also, there are a couple of relevant design requirements in the FARs that apply to aileron deflections. At the design cruise speed, V_c (which is discussed in the airspeed limitations subsection of this chapter), the structure must be designed for the aileron deflection required to produce the same roll rate you get at V_a with full aileron deflection. At the design dive speed, V_d (also subsequently discussed), the aircraft must be designed for one third this roll rate. Aileron deflections produce twisting loads on the wing and you must be especially careful when applying aileron at speeds faster than V_a.

Also bear in mind that the limit load factors apply only to symmetric maneuvers. For all the aileron input cases discussed above, the aircraft is assumed to be pulling up with only two thirds of its positive limit load factor; thus, for example, an aerobatic airplane designed to 6G in a symmetric pull-up at V_a need only be designed to a 4G rolling pull-up with the ailerons fully deflected. Pilots operating handbooks offer little guidance in these matters, but suffice it to say that pilots should avoid combining large aileron deflections together with high G and/or speeds faster than V_a.

The right side of the maneuvering envelope is limited by V_d, the *design dive speed*. This speed is selected by the aircraft designer (subject to a certain minimum). Some possible consequences of flying beyond this speed (other than overloading the aircraft) are *flutter*, *divergence*, and *control reversal*.

Flutter. Unstable vibration of the aircraft might occur without warning and can involve the whole aircraft or just a component, such as a control surface. Flutter, when it occurs, is often catastrophic.

Divergence. Wing divergence involves an uncontrolled twisting and bending of the wing, leading edge, and wingtips. It usually results in the sudden departure of wings.

Control reversal. Aileron deflection results in wing-twisting loads. The resulting wing twist is an effective aileron input, the direction being opposite to that of original aileron input. As the speed increases, this effect becomes stronger, and at a certain speed the wing twist might win out over the ailerons, so that the aircraft will roll in a direction opposite from that intended.

Pilots should also be aware that the regulations allow the lower right corner of the maneuvering envelope to be cut somewhat (as is the case in our sample envelope). This could be critical in negative-G high-speed maneuvers. Refer to the flight envelope for your specific aircraft.

While the maneuvering envelope defines the loads that are to be expected in symmetric maneuvers, the *gust envelope* defines loads that might arise when the aircraft encounters a gust. Because for a given gust strength the gust load increases with the airspeed, the regulations define different gust strengths at different airspeeds. A *design cruising speed*, V_c, is defined at which the aircraft must be designed for gusts of 50 feet per second. At V_d, the structure need only be designed for gusts of half this strength. This is why you should avoid exceeding V_c (actually, V_{no}, which is explained in the next subsection, airspeed limitations) in gusty weather.

As mentioned before, the flight envelope contains the maneuvering envelope and the gust envelope. It is not, however, an envelope of combined maneuvering and gust loads; thus, for example, there are no design requirements addressing gust encounters during pull-ups. This is something pilots might wish to consider before performing aggressive pull-ups or pushovers in gusty weather. Regardless of the gustiness of the weather, during maneuvers, remain within the maneuvering envelope, even if it is exceeded by the gust envelope.

Airspeed limitations and safety factors

Of the design airspeeds V_a, V_c, and V_d, only V_a finds its way into the pilot's operating handbook. Instead of V_c and V_d the airspeed limits V_{ne} and V_{no} are provided to the pilot as operating limitations. V_{ne}, the *never-exceed speed*, is set just under V_d by a small safety factor. As the name suggests, this is the speed that should never be intentionally exceeded under any circumstances. Observe this speed; don't try to take advantage of the safety factor.

However, should you ever find yourself inadvertently exceeding V_{ne}, for example, as the result of a botched maneuver, remember that in the pullout the aircraft is still subject to structural limits. Resist the temptation to pull as hard as possible on the elevator; it does little good to get back to slower than V_{ne} if you no longer have the benefit of the wings. The safe recovery technique is to immediately retard the throttle, level the wings, and *very slowly* pull on the stick to bring the nose to the horizon.

The design cruise speed, V_c, shows up in the pilot's manual as V_{no}, the *maximum structural cruising speed*. V_{no} is equal to or slightly less than V_c. To you, the pilot, V_{no} is the speed that should not be exceeded, except in smooth conditions. Also, keep in mind the discussion above regarding the aileron deflections at V_c and beware of excessive aileron inputs when indicating faster than V_{no} (full aileron deflections should not be made when faster than V_a).

The airspeeds V_a, V_{no}, and V_{ne} are established as operating limitations and the FARs require you to observe them. V_{ne} is marked on the airspeed indicator with a radial red line. A yellow arc is painted on the ASI from V_{ne} to V_{no}. Below this, a green arc extends down to the flaps-up stall speed. The point where the green and yellow arcs meet is V_{no}, which is not the maneuvering speed V_a, as many pilots seem to believe. Because V_{no} is generally faster than V_a, this is an important distinction. The maneuvering speed should be listed on a placard on the instrument panel near the airspeed indicator.

The loading conditions we have discussed are *limit loads*, meaning the maximum loads to be observed in service. The aircraft must actually be designed to *ultimate loads*, which are the limit loads plus a safety factor of (in most cases) 50 percent. The safety factor is intended to cover variations in material properties and uncertainties in load assumptions; it is not intended to cover the case of pilots pushing their luck.

You should know that although at each loading condition the airframe is required to support the ultimate load, the FAA allows local structural failures (popped rivets, broken ribs, and the like) to occur as long as they occur above limit load and as long as the overall structure remains intact. Also, permanent deformations do not count as structural failure and might occur anywhere above the limit load. So even if the aircraft holds together, exceeding limit load might, for example, bend the wing spar, leading to the grounding of the aircraft.

One scary scenario is the pilot who routinely exceeds the aircraft's limitations, feeling entitled to take advantage of safety factors used in the design. After all, because the airplane hasn't shown the slightest symptoms of imminent disintegration, obviously the airplane is overdesigned, right? Unfortunately for this pilot—or too often the unsuspecting fellow pilot who flies the aircraft next—the airframe might be damaged in ways that might escape notice, and imposing large repetitive loads on it increases the risk of a structural fatigue failure. All materials "remember" loads to some extent, but metal is especially susceptible to the cumulative effect of repeated high loads (consult your aircraft manual and the manufacturer). Fatigue failures can occur at loads well below what would be required to break the structure in a single application. Take airspeed and structural limitations seriously.

THE IMPORTANCE OF ANGLE OF ATTACK

Of the two concerns most often expressed by pilots interested in learning to fly aerobatics, we have chiefly dealt with the possibility of falling apart—experiencing structural failure. Now let's address the possibility of falling out of the sky—finding ourselves in a stall, which we all know can even develop into a dreaded spin. How can it come about that lift is no longer generated by the wing and the airplane "stops flying?" It all has to do with only one thing: *angle of attack*.

To review what you learned in ground school: The airflow parallel and opposite to the direction of flight is the *relative wind*. Within this airflow is the wing, generating lift. The imaginary line between the leading edge and the trailing edge of the wing is the *chord line*. The angle between the relative wind and the chord line is the *angle of attack*. The maximum angle of attack at which the wing is capable of generating sufficient lift to sustain flight is the *critical angle of attack*. When the angle of attack exceeds the critical angle of attack the aircraft *stalls* (FIG. 2-7).

What is also simple and very important for all pilots and especially aerobatic pilots to understand is that the airplane can stall at any speed because the critical angle of attack can be exceeded at any speed. All it takes is to change the aircraft's pitch at a faster rate than with which the change in the relative wind can keep up. In a normal pull-up, as the pitch angle increases, the load factor increases, curving the flight path upward. The relative wind thus changes direction and reduces the angle of attack. But in a very abrupt pull-up, the angle of attack can increase so quickly to the stalling angle that the relative wind cannot keep pace. A rapid yank on the stick can do it.

Look at FIG. 2-8. When the pilot of the Staudacher flying at 150 knots (faster than the stall speed) quickly snatches the stick aft, the aircraft is still moving parallel and opposite the original relative wind. Because the pitch was changed so rapidly, the relative wind could not change fast enough to keep up with the wing. Consequently, the critical angle of attack has been exceeded and the Staudacher stalls at 150 knots. If the pilot had changed the pitch at a lower rate, the relative wind would have had time to readjust to the changing pitch angle, the aircraft would not have exceeded the critical angle of attack, and the aircraft would not have stalled.

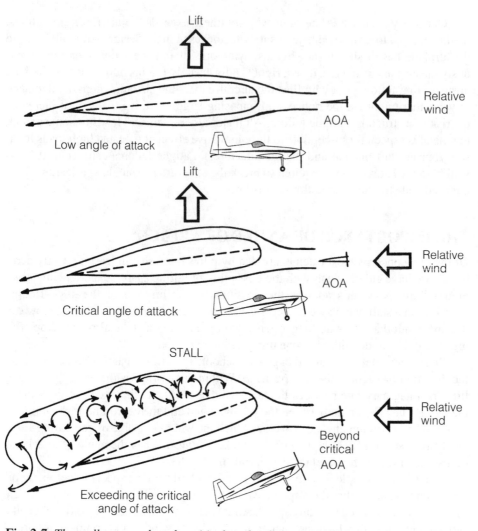

Fig. 2-7. *The stall occurs when the critical angle of attack (AOA) is exceeded.*

This tells us that any aircraft's published stall speeds apply only to a particular set of conditions (power-off straight-and-level). The published stall speeds are very useful and important reference speeds for nonaerobatic, slow-speed operations, but is only a very small part of the story of the stall.

To illustrate the point, consider this: Low-powered aircraft lose so much speed by the time they reach the top of a loop, that the ASI reads practically zero; why doesn't the airplane stall? It doesn't because the pilot adjusted the pitch change at a rate that has enabled the relative wind to keep pace and the critical angle of attack is never exceeded (FIG. 2-9).

The aerobatic pilot will find it useful to always know where the relative wind is coming from and to think in terms of angle of attack.

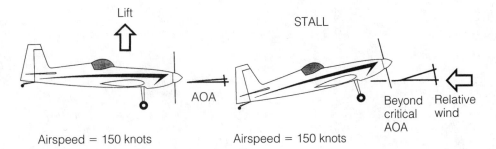

Fig. 2-8. *The accelerated stall. Abrupt aft stick pitches the aircraft up—the relative wind is unable to keep pace with the pitch change—the critical angle of attack (AOA) is exceeded and the aircraft stalls.*

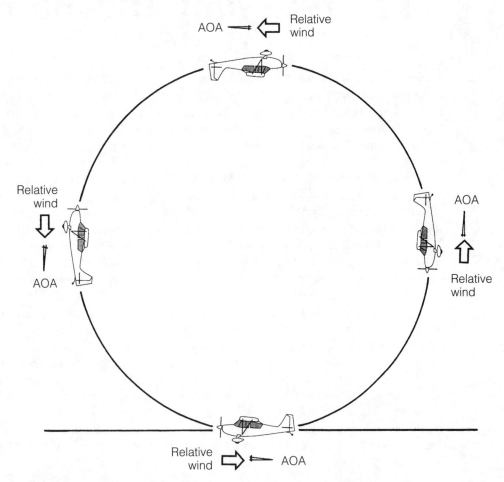

Fig. 2-9. *Why doesn't the aircraft stall in a loop? Because the relative wind keeps up with the flight path change and the critical angle of attack (AOA) is not exceeded.*

The first loop the loop

Adolphe Pegoud, the intrepid Frenchman who was the first to jump out of an airplane with a parachute and got the idea for experimenting with aerobatics when he witnessed the antics of his abandoned Bleriot, has been thought by some to have performed the first-ever loop, or, as it was commonly referred to back then, the "loop the loop." Pegoud's approach to aerobatic flight was systematic and analytical. During the summer of 1913 through a series of split S maneuvers, he was, indeed, steadily working up to the full loop. But he was beaten to it by a young Lieutenant of the Russian Imperial Air Service, Pyotr Nikolaevich Nesterov, by just 12 days.

Nesterov had been flying for about two years, and was known as an aggressive innovator, when on September 9, 1913, he fired up his Nieuport IV monoplane and climbed to 3,000 feet in the vicinity of Kiev in the Ukraine. He cut the engine and dived, gathering speed. He had to literally shut it off by shutting off the fuel because his Nieuport, like many aircraft of the day, had no throttle and the engine was either full on or shut off depending on fuel flow.

At 1,900 feet, Nesterov pulled hard on the control column, switched the engine back on for extra power, and gracefully arched all the way around in the vertical plane to complete the first ever "loop the loop." He then cut the engine once again and glided down to a smooth landing and was promptly jailed for endangering state property. It took his superiors 10 whole days to figure out the significance of his achievement, whereupon he was promptly released and promoted.

Nesterov was dead within a year, having deliberately rammed an intruding Austrian military aircraft in the opening days of World War I. Today he is remembered by the ornate and much-prized Nesterov Cup awarded to the winning team of the world aerobatic championships.

The origin of the term "loop the loop" is an interesting morsel of aerobatic history, and it has nothing to do with flying. It is an American term, borrowed from an amusement park ride. It seems that already in the early 1900s, Americans were keen to be titillated in even more imaginative ways. One such thrill was a roller coaster ride that included a full circle in the vertical plane. To ride the circle was to "loop the loop."

FORCES SPECIFIC TO PROPELLER-DRIVEN AIRCRAFT

In addition to being concerned with general loads, limits, and stall characteristics, the pilot of the propeller-driven airplane has to consider the effects of a separate set of forces created by the rapidly rotating propeller. Think about it for a minute. A propeller turning at 2,400 RPM turns 40 times a second. If you ever picked up a propeller you know how heavy it is; there are some serious forces at work bolted to the nose. Again, you will be familiar with these forces from ground school, and they do have an effect on the subtleties of aerobatics that you will notice as you gain experience, so it is worth briefly reviewing them. Note that the definitions below apply to aircraft with propellers rotating to the right. For aircraft with propellers turning to the left, the directional effects and direction of corrective action will be exactly the opposite.

P factor. A propeller is simply an airfoil attached to a common hub. In level flight, the ascending and descending blades of the propeller have equal angles of attack, producing uniform lift. As the aircraft enters a climb, the descending blade meets the air at a greater angle of attack than the ascending blade and as a result produces more thrust. This increase in thrust produces a left turning tendency known as *P factor*, which must be countered by right rudder.

Torque. For every action there is an equal and opposite reaction (Newton's first law). A propeller rotating in one direction will want to cause the aircraft to rotate in the other direction. This effect is known as *torque*. In straight and level cruise speed, it is compensated for by the angle at which the engine is installed in the aircraft. Because the installation compensates for torque only at cruise speed, the aircraft wants to rotate to the left at slower airspeeds unless controlled by aileron. Torque is most noticeable during hammerheads and slow-speed pushovers at the top of a vertical line.

Slipstream. The rotating propeller forces air backward around the aircraft in a corkscrew path. In cruise flight, the slipstream is sufficiently elongated to not disturb the fuselage. As the aircraft slows, the slipstream tightens and begins to flow against the aft left portion of the fuselage, causing the nose to move to the left. The slipstream's effect is most noticeable in the slow-speed portion of vertical maneuvers. Slipstream is compensated for by opposite rudder.

Gyroscopic precession. The rapidly spinning disk of the propeller possesses a rigidity in space and is in essence a giant gyroscope, similar to the tiny spinning gyros in attitude and heading indicators in the cockpit. The faster the rotation, the greater the rigidity. At typical propeller rotation speeds and airspeeds, the propeller's gyroscopic effect is not very noticeable. At low speeds combined with rapid pitch changes, the effect can be dramatic and has led to the creation of a whole class of gyroscopic effect-induced aerobatic maneuvers such as the Lomcovak and the Bessenyei.

According to the laws of physics, when a force is applied to a spinning object, the effect will be felt 90° ahead of the point of application in the plane of rotation; thus, when the nose is pulled up rapidly, the force applied to the rapidly rotating propeller disc is on the bottom. The effect will be felt 90° ahead of this point in the plane of rotation; the effect will be as if a force were pushing the left side of the prop to the right; it is corrected by left rudder.

When the stick is pushed forward (as in entering a negative-G maneuver) the force applied to the propeller disc is at the top. The effect is felt on the right side, pushing the prop to the left, to be corrected by right rudder.

This pretty much takes care of the theory you need to know while you are doing aerobatics. As we discuss the individual maneuvers, we will return to these concepts to see how they apply.

ACCELERATION AND THE HUMAN BODY

We have seen how acceleration loads affect the airplane, but they also have a significant and potentially incapacitating effect on the human body that need to be equally understood. High Gs impose as punishing a load on the body as they do on the airplane. A person weighing 180 pounds under normal 1-G conditions weighs 1,440 pounds in an 8-G maneuver 8×18°. It is a real challenge for the cardiovascular system to keep various extremities of the body, including the brain, supplied with the appropriate amount of blood, and at some point the body will no longer be able to cope, ultimately resulting in a loss of consciousness. Individuals are affected to different degrees by G forces, and G tolerance increases with regular exposure to G loads.

Most pilots rarely encounter the impairing effects of G forces during basic aerobatics, but everyone flying advanced maneuvers with higher G loads is eventually affected. Depending upon the maneuvers flown, the body experiences a series of symptoms as G forces increase, enabling the pilot to recognize and manage G effect. Three basic types of G effect are *grayout-blackout*, *redout*, and *G-Loc*.

Grayout-blackout

Grayout and blackout occur in positive G maneuvers, such as a high-G pull-up into a loop. Blackout is the more serious second stage of grayout. The human body gives ample warning of the onset of grayout. First comes tunnel vision, then you see everything in black and white, gradually—from the periphery inward—the world begins to turn gray, darkening as the Gs are sustained. Eventually the world goes black, though for a while you can still hear before you drift into unconsciousness. Ease off the Gs, which you should do well before the onset of blackout, and the process is reversed as your hearing and vision return to normal.

During grayout and blackout the direction of the G force is away from your brain, drawing blood away from it. The heart is unable to pump a sufficient amount of blood against this force, and at approximately 5Gs in the average person, blood flow to the brain ceases all together. As the brain is progressively deprived of the blood's oxygen, grayout and blackout sets in. At the point when blood flow ceases completely, approximately 3–5 seconds of consciousness remains as the brain exhausts its last reserves of oxygen.

Redout

Redout is the opposite of grayout and blackout, and is encountered in negative-G maneuvers. In this case the direction of the G force is toward the brain, causing blood to rush toward it. The heart is unable to sufficiently regulate blood flow to prevent excess blood from flooding the brain's blood vessels and causing redout. Redout rarely affects pilots who regularly experience negative Gs and are in excellent physical shape. Redout can cause the rupture of blood vessels, such as the hairline vessels in the eye or the eyelid that will leave red spots for a day or two. Your bottom eyelid might float up over the eyeball under negative G; this will also cause redout symptoms that might hinder your vision until the negative Gs are relieved. The reduction of Gs reverses the phenomenon, and vision and hearing return to normal.

G-Loc

G-Loc, which stands for *G-induced loss of consciousness*, is a nastier phenomenon than grayout or redout. It results in an instant loss of consciousness with no warning. It occurs when high Gs (usually more than 5Gs) are pulled rapidly and are then sustained for a long time (usually 5 seconds or more). Loss of consciousness can last up to 15 seconds or more.

The good news is that G-Loc is rarely a concern for basic aerobatics because its preconditions are not produced during basic maneuvers, and with the exception of the most advanced types, civilian aerobatic aircraft are generally incapable of sustaining high Gs long enough to cause it.

A condition under which G-Loc might occur is during a transition from high negative Gs to high positive Gs. It is vitally important to work up to such maneuvers gradually, practice appropriate anti-G straining (explained in the next subsection of this chapter), and generally be in top physical shape.

G-Loc was discovered and is primarily encountered by fighter pilots whose maneuvers in their modern, super-agile jets regularly create the required conditions. In the aerobatic community, G-Loc has recently become a factor for aerobatic competition pilots at the highest performance level because of the great increases in competition aircraft performance; however, as high-performance dual aircraft become more readily available at aerobatic flight schools and to private owners, it is worthwhile for all of us to understand G-Loc.

Anti-G straining

A useful technique against the effects of positive G is to learn to tense the abdominal and chest muscles through the breathing cycle during the maneuver. A quick intake of breath should be followed by tensing the abdominal and chest muscles while slowly exhaling over about 3–4 seconds. The breathing cycle is then repeated. Without going into medical details, the general idea is to make it harder for the blood to flow away from the brain. Recent research has shown that grunting and holding your breath while you tense your muscles is less effective than the described breathing technique.

Anti-G straining makes it easier to tolerate G forces at all levels, so it is useful for all aerobatic pilots. Experiment on your dual instructional flights for the first few times and you will soon get the hang of it. Experienced pilots who properly apply the technique can increase their G tolerance by up to 3Gs.

Building G tolerance

Another way to increase G tolerance is by simply experiencing G loads on a frequent and regular basis. If you consistently fly aerobatics two or three times a week, you will soon find that you are routinely able to withstand G loads that in the beginning would have caused grayout or worse.

If you stop flying aerobatics for a while, you will lose the G tolerance you had built up, and will have to rebuild it. Competition pilots start to sense a lowering of G tolerance if they lay off regular practice for as little as a week.

Fig. 2-10. *Pushing the envelope.*

Know when to stop

Regardless of your ability to tolerate Gs, your body takes a pounding every time you fly an aerobatic session and eventually it will say, "Enough!" Be sure to recognize the symptoms and head for the barn before you become too exhausted or spaced out to comfortably fly the airplane. Excessive fatigue, increasing discomfort during repeat maneuvers, weird respiratory sensations such as feeling out of breath, increasing difficulties with orientation in a maneuver, and seeing an occasional star or two in broad daylight are all signs to call it a day.

3
Preflight and airborne preparations

Y OU HAVE SELECTED AN AEROBATIC SCHOOL AND AIRPLANE, YOU AND
your instructor have covered the aerodynamics and structural limits important to
aerobatics, and you are ready to see what it is all about. Before you take off on your
first lesson, the first thing you will have to do, as you would on any flight, is preflight
the aircraft.

THE AEROBATIC PREFLIGHT

In many respects the aerobatic preflight is no different from a regular preflight, and in
all cases you should follow the manufacturer's preflight procedures in the aircraft operating manual; however, a number of preflight items common to most aerobatic aircraft and rarely checked otherwise can tell you a lot about whether or not to launch for
an aerobatic session. You should especially look for signs of overstress, and take extra
care with some regular checklist items because of the structurally demanding flying
done by the airplane. The items below are not a comprehensive preflight checklist, but
a partial listing of items especially applicable for aerobatic flight.

Accelerometer (G meter). Check the accelerometer for the positive and negative
Gs it registers from the previous flight to make sure they are not beyond limits. If ei-
ther one is beyond its limit, do not fly before consulting an appropriate mechanic. If the
aircraft is used by several people, talk to the individual who last flew the aircraft.
When satisfied that G limits have not been exceeded, reset the G meter to record the
Gs you will pull on your flight.

Control movement. Upon entering the cockpit, carefully check the controls for
freedom of movement in all directions. Listen for any binding in the wings and fuse-

Fig. 3-1. *A good aerobatic preflight is essential. The pilot is looking for stress wrinkles on the wing skin.*

lage. Binding might occur because of a loose item caught in the control system or structural distortion.

Parachute. Make sure the parachute is current (within the timeframe beyond which it has to be repacked). Check to see if the rip cord is in position, the harness and buckles are in good condition, and make sure there are no visible signs of damage. Always store your parachute in a dry, rodent-safe place.

Seat belt attach points. Check the attach points for wear and tear. The seat belt attach points of an aerobatic aircraft can wear out because of the constant twisting forces acting on the attach bolt.

Loose items in cockpit. Make absolutely sure that there are no loose items in the cockpit. The smallest loose screw can cause the controls to jam if it slips in the wrong place. The litany of hair-raising scrapes with controls jammed by loose items is endless.

Loose items in fuselage. Carefully look in the fuselage for any loose items and tap the underside of the fuselage vigorously while listening for any loose items rattling about. Always check around the tail area of the fuselage because this is where loose items tend to gravitate, especially in taildraggers. Broken fuselage frames, popped rivets, screws, bolts, pens, coins, forgotten pliers, screwdrivers, clothing, and sleeping cats are among the many things discovered in the fuselage of aerobatic aircraft, not always in time. In Decathlons, the belly frames tend to break from time to time and slide into the tail.

Fig. 3-2. *Loose objects found in the fuselage on aerobatic preflights.*

Wing condition. Check for general condition and surface wrinkles. Look for any sign of structural damage. Fabric-covered aircraft reveal structural deficiencies earlier than metal-covered aircraft. On single-seat Pitts Specials, watch for loose rib nails through the fabric, a sure sign of having been flown hard.

Ailerons. Check for general condition and security. Check for excessive wear of the control hinges and attach points, which could cause flutter.

Shovels. The ailerons of most aerobatic aircraft are fitted with counterbalances (nicknamed shovels) that reduce control pressure in roll maneuvers. The shovels consist of an arm bolted to the aileron bracket and a flat plate attached to the arm. Great stress is exerted in flight on the plate and the arm where it attaches to the plate. Check this area carefully for cracks and any other sign of stress. The shovels on Pitts Specials are a weak point that needs constant attention.

Fuel and oil caps. Check for tightness. In inverted flight, a loose cap can siphon off fuel or oil at an incredible rate.

Propeller hub. Check the propeller hub for cracks on aircraft equipped with constant-speed propellers. Oil weeping onto the front of the propeller blades is a sign of a cracked hub. Loss of propeller control can result.

Fuselage. Check for wrinkles, loose rivets, and any other sign of wear or structural damage.

Bracing wires. Most biplanes and some monoplanes are equipped with bracing wires that are an integral structural part of the wing and tail surfaces. Check the bracing wires for nicks that are usually caused by rocks thrown up by the propeller during taxi operations. A nick weakens the bracing wire and is structural damage: *Do not fly the airplane until the nicked bracing wire is replaced.*

PILOT'S POCKETS PREFLIGHT

It might seem a bit much to have a separate section for the pilot's pockets, but that is where a majority of the loose items that worm their way into the fuselage come from. So it is absolutely imperative to *empty pockets of change, pens, and all other loose items prior to flight.*

IN-FLIGHT PREAEROBATIC PREPARATION

Well, how about that? You are in the air in an aerobatic airplane, about to begin some of the most exciting flight training you will ever experience. Now is the time to check out your practice area.

The aerobatic practice area

Generally there is really no such thing as an official aerobatic practice area. Legally you may perform aerobatics anywhere where you are not prohibited from doing so by the FARs; however, there might be local regulations in effect regarding the airspace in the vicinity of your airport that you need to observe. Often there are noise-sensitive areas to avoid. Some flight schools have preferred practice areas, selected based upon experience and perhaps informal consultation with a variety of authorities. The best rule of thumb is to find out about and accommodate local practices.

If you fly out of an airport where there are no customary practice areas, no local habits to observe or airspace issues to worry about, then the selection of your practice area is entirely up to you and your instructor. In this case, your best rule of thumb is a good dose of common sense.

Consider the locations of airports and off-airway VFR traffic flows in the vicinity and select an area that is most likely to be less traveled. Take into consideration any landmarks you might wish to use as reference points, such as prominent hilltops or mountain peaks and straight stretches of road. And it never hurts to have a few suitable emergency landing areas within reach. Having arrived in and briefly reviewed with your instructor the practice area that has been selected by the instructor, it is time to do your final preaerobatic checks.

In-flight preaerobatic checks

The in-flight preaerobatic checks are the last chance to make sure that everything is in order, and an opportunity to make any special settings specifically required for aerobatic flight. On the way to the practice area, you might have had a window open, a water bottle loose in the cockpit, and the power set differently from the requirements of aerobatic flight. If your aircraft is equipped with gyroscopic instruments, such as an artificial horizon or a directional gyro, you need to cage them before aerobatics to prevent them from tumbling. Now is the time to clean everything up.

A typical preaerobatic check might contain the following items (consult your aircraft operating manual for all checks required by your specific aircraft):

- ❑ All loose items stowed and secure
- ❑ Seat belts and aerobatic harness fastened and tight
- ❑ Canopy locked and secure
- ❑ Gyros caged
- ❑ Engine instruments in the green
- ❑ Flaps and gear up
- ❑ Power set for aerobatic flight

Checks complete, you are committed to aerobatic flight, and it is time to do your clearing turns.

Clearing turns

The purpose of clearing turns is to ensure that no other traffic is about to intrude on your aerobatic session. It is the pilot's responsibility to maintain separation from other aircraft throughout the aerobatic session—as on all other VFR flights—and the clearing turns are a part of this process. In straight-and-level flight, many sectors of your field of vision are blocked. It is essential to carefully check all quadrants for traffic because during an aerobatic session you are rapidly moving through a big piece of sky in all dimensions.

A 90° turn to the left, followed by a 180° turn to the right followed by another 90° turn to the left will cover all 360° around you and bring you back on your original course. During the turn, carefully and systematically scan the airspace vertically as well as horizontally.

The clearing turn should be done in a wide arc in a shallow bank. Some pilots make the mistake of doing clearing turns snappily at near-vertical bank angles, which might warm them up for the session to come but greatly diminishes the maneuver's value as a clearing exercise.

It is a good idea throughout an aerobatic session to do additional clearing turns from time to time. A good opportunity is when you want a short break from the maneuvers.

Clearing turns complete, sky clear, and it is time at last to get down to business.

4
Stalls

IN THE STALL, THE AIRCRAFT'S WING EXCEEDS THE ANGLE OF ATTACK beyond which it is no longer able to generate lift (the *critical angle of attack*). In the absence of lift, the aircraft's nose pitches down until the wing's angle of attack is reduced below the critical angle of attack and lift is once again generated. This chapter covers only upright stalls entered from upright flight. Inverted stalls are addressed under the chapter on inverted flight.

It is important for the aerobatic pilot to understand the stall in great detail. It is the lower limit of the performance envelope and in many aerobatic maneuvers the aircraft approaches it closely. Overstepping it results in an incomplete maneuver, and might also be unsafe depending upon altitude and several other factors.

UNDERSTANDING IT

Two key factors need to be understood regarding the stall: relative wind and angle of attack (FIG. 4-1). As briefly discussed in chapter 2, relative wind is the airflow opposite and parallel to the direction of flight; angle of attack is the angle between the wing's chord line—the imaginary line from the leading edge to the trailing edge—and the relative wind.

Smooth airflow over the surface of the wing generates lift. Generally, the higher the angle of attack while the airflow remains smooth, the greater the lift generated; however, there comes a point where the angle of attack is so high that the airflow can no longer adhere to and flow smoothly over the wing's surface. The airflow separates from the wing surface and becomes turbulent. Lift is no longer generated and the aircraft stalls. The angle of attack at which the aircraft stalls is the critical angle of attack.

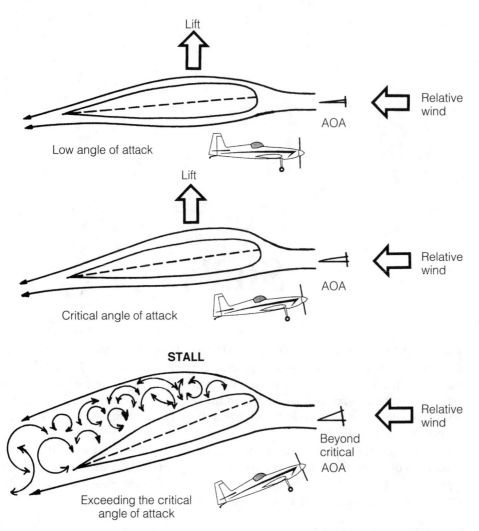

Fig. 4-1. *The stall occurs when the critical angle of attack (AOA) is exceeded.*

You can stall at any speed. To stall an aircraft, you can bring up the nose slowly, in which case you bleed off the airspeed and the aircraft will stall at a slow airspeed in the range of the published power-on and power-off stall speeds. Or, with an abrupt control movement, you can pitch up the aircraft at such a rapid rate that, although inertia is still carrying the aircraft in its original direction, the direction of the opposite and parallel relative wind remains unchanged, you have exceeded the critical angle of attack. You will stall, even though the airspeed indicator is way above any of the published stall speeds. This is known as the *accelerated stall*. It is sometimes said that the pitch change is too rapid for the relative wind to readjust to it; hence, the critical angle of attack is exceeded. Never forget that an aircraft can stall at any speed. Always know the direction that the relative wind is coming from.

Recovering from the stall. The key to recovering from the stall is to reestablish an angle of attack below the critical angle of attack. This is accomplished by pitching the nose down. Most aircraft have, by design, an intrinsic tendency to recover from the stall. The *center of lift* (the aircraft's point of equilibrium) is between the aircraft's center of gravity and the horizontal tail surface. The center of lift is closer to the c.g. than it is to the tail (the moment is less); therefore, under conditions of lift, the countervailing downward force at the c.g. is greater than the downward force at the tail, and given the proportionally different moments, there is equilibrium. When lift ceases, the aircraft slows and the tail force also diminishes, causing the aircraft to pitch down around the c.g.

Aircraft controllability and the stall. Aircraft designers take great care to make their aircraft as well behaved in the stall as possible. Not all of the wing stalls at the same time, and controlling the airframe's stall sequence is the trick to creating benign stall handling characteristics. The objective is to keep the outer section of the wing flying as long as possible to retain aileron control (FIG. 4-2).

This is accomplished by designing a slight outward twist into the wing. As a result, the outer section of the wing has a lower angle of attack, and will remain flying when the inner section has already stalled. Because the airflow always separates from the aftmost part of a wing first and works its way forward, the stall pattern of a wing

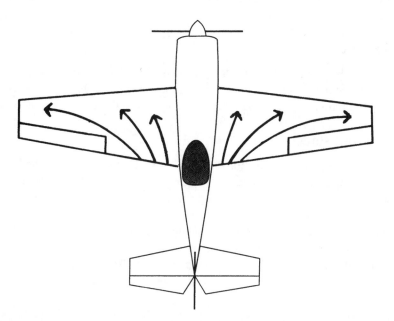

Fig. 4-2. *Typical wing stall sequence: aft inboard first, then forward and outward. Note, aileron is affected last.*

designed with a decreasing angle of attack outward is: aft inboard, then forward, and lastly outward; thus, aileron control is available until the final onset of the full stall.

When the angle of attack is decreased to recover from the stall, the sequence is reversed. The outer portion of the wing resumes flying first, instantly providing aileron control.

Another desirable design feature is some form of stall warning inherent in the behavior of the airframe just prior to stalling. This is accomplished by the *prestall buffet*. Most horizontal stabilizers are attached to the airframe in a position that puts them in the path of the turbulent, separated airflow thrown off by the aft inboard portion of the wing when the rest of the wing is still generating enough lift to keep the wing flying. It is this turbulence shaking the horizontal stabilizer that we feel when we sense prestall buffet.

The stick as angle of attack indicator. One of the more coveted devices in the cockpits of many military airplanes and civilian turbojet aircraft is an angle of attack indicator. But there is also such a device in all aerobatic aircraft (indeed, all aircraft) and it is none other than the stick. In upright flight, any time the stick is aft of neutral and remains there, you are on the way toward a stall. The farther back the stick is, the closer you are to stalling. So, if at any time—in whatever unusual upright attitude and at any speed—you are concerned about being too near the critical angle of attack, check the position of the stick. Move it to neutral or farther forward and you will remain below the critical angle of attack.

Similarly, in inverted flight, any time the stick is forward of inverted neutral, a stall is imminent; move it to neutral or farther aft and you will remain below the critical angle of attack.

FLYING IT

In addition to thoroughly reviewing and practicing the standard power-off and power-on stalls first encountered in basic flight training, the aspiring aerobatic pilot is well served by in-depth additional training in accelerated stalls and stalls at a variety of bank angles.

Power-off stalls

At cruise airspeed retard the throttle to idle. As the aircraft slows, the lift will also decrease. To maintain straight and level flight you must, therefore, increase the angle of attack by gradually applying up elevator (aft stick). Eventually you exceed the critical angle of attack and the aircraft stalls.

To recover, lower the angle of attack below the critical angle of attack by down elevator (forward stick) and you will be flying again as enough speed is picked up to restore smooth airflow over the wing to generate lift. To complete the power-off stall maneuver, pull out into straight and level. Be careful not to pull out too abruptly, which could cause a secondary stall.

While normally we add power after the nose drops to recover from the power-off stall, note that this has nothing to do with recovering, other than expeditiously providing flying speed.

In certain aircraft, it might happen that a wing will drop when you stall. Most novices' instinctive reaction is to counter instantly with opposite aileron and it is *totally wrong*; opposite aileron in a stalled condition is the perfect recipe for a *spin* because aileron adds drag to the inner (down) wing, increasing the autorotational forces. The only correct recovery action in a stall is *forward stick* to unstall the wing. If a wing drops immediately, pick it up with opposite rudder while applying forward stick to unstall the wing. If a spin has developed, apply opposite rudder first to stop the rotation followed by forward stick to unstall the wing (*see* chapter 14, regarding spins). The aileron should be kept neutral throughout the maneuver.

Power-on stalls

Reduce power to slow the aircraft down. Raise the nose. As the aircraft slows and lift diminishes, you gradually restore power to maintain straight and level while gradually continuing to bring the nose up. Because of P-factor and slipstream effect, the nose of an aircraft with a propeller rotating to the right eventually wants to wander to the left, gradually requiring a good dose of right rudder to maintain straightline flight. When the critical angle of attack is exceeded, the aircraft stalls and the nose drops.

To recover, reduce power, apply down elevator (forward stick), and pull out of the dive, restoring power as you do. If a wing drops, pick it up with *opposite rudder*. Maintain neutral aileron throughout the maneuver.

Accelerated stalls

Practice the accelerated stall by getting the airplane reasonably slowed down but well above the published stall speed, briskly apply up elevator (aft stick), and hang on. If your aft stick movement was sufficiently brisk, you will experience a stall that is generally similar to the normal stalls but happens much faster. To recover, apply down elevator (forward stick) to reduce the angle of attack below the critical angle of attack and you will soon be flying again.

Banked stalls

Any stall, power-off, power-on, and accelerated, should be performed not only from straight-and-level flight but also from various bank angles, duplicating the conditions under which inadvertent stalls might be encountered during the unusual attitudes of aerobatic flight. To practice various banked stalls, set them up exactly as you would set them up straight and level, except keep the aircraft in the appropriate bank.

When you are banking, you are also turning in the direction of the low wing; therefore, the high wing travels faster and the low wing might stall first, creating a tendency to roll over the top. Disconcerting as this might look to the novice, the initial recovery technique is exactly the same as in stalls in which a wing has dropped: stick forward to unstall the wing and opposite rudder to pick it up.

Under no circumstances should you immediately counter the roll with opposite aileron; it could lead to a spin.

Practice stalls of all kinds diligently with your instructor and feel completely comfortable with them before moving on to aerobatic maneuvers. With a little training and effort, stalls of all types will become routine, and that is what you as an aerobatic pilot want them to be.

COMMON ERRORS

Several common errors occur during stall training, all easily overcome with practice.

The stall is not entered in coordinated flight. If the stall is entered uncoordinated, a wing might drop because one wing will have a higher angle of attack (because of being in uncoordinated flight) and it will stall first. Apply forward stick—to unstall the wing—followed by opposite rudder—to pick it up—and reestablish the heading.

Some aircraft do not break into a stall when full aft elevator is applied gradually, they just mush along at a high sink rate with wings level. You can cure this problem by waiting until you are 5 knots above the stall and applying brisk, smooth aft stick. The nose will rise smartly and the aircraft will break cleanly into a stall.

The aircraft climbs as it approaches the stall. This is just sloppy flying cured by paying attention and practicing a lot.

IF THINGS GO WRONG

Several things can go wrong in a stall to the extent of possibly compromising safety, especially at low altitude. Most common are the inadvertent secondary stall and the inadvertent spin.

Inadvertent secondary stall. If your pullout is too abrupt, you might induce an accelerated stall immediately after recovery. Prevent it by pulling out of the original stall quickly but not abruptly. If you do experience a secondary stall, don't panic; recover from it as you should have from the original stall.

Inadvertent spin. An inadvertent upright spin might develop if you drop a wing in the stall. Consult your aircraft manual for the appropriate spin recovery technique for your particular aircraft and receive dual spin instruction from a qualified instructor. The standard spin recovery technique, covered in great detail in chapter 14, will generally work:

- Power off
- Opposite rudder to stop the rotation
- Forward stick
- When the spin stops, neutralize all controls and pull out of the dive

Under no circumstances respond to a spin or a wing drop with opposite aileron. The aileron should be kept neutral throughout the recovery.

(Chapter 14 also presents the Beggs/Mueller emergency recovery technique that works in most aircraft.)

Jimmy Doolittle
and the outside loop

By the time the late Jimmy Doolittle led the famous B-25 raid on Tokyo during World War II, he was already a living aviation legend, a reputation he earned as a glamorous racing and stunt pilot during the golden age of American aviation. Tales of the daredevil in him often overshadow the fact that Doolittle also held a graduate degree in aeronautics from the Massachusetts Institute of Technology and was a pioneer in meticulously applying principles of science to test flying. It was this careful and measured approach that enabled him to develop the maneuvers that, over an airshow crowd seemed, like sheer daredevilry.

A maneuver yet to be conquered as Doolittle came into his element on the airshow circuit was the outside loop. Negative, downward, half loops had been flown before, first at the Royal Aircraft Establishment at Farnborough (hence the maneuver was known for awhile as the *English bunt*). But all attempts to push up all the way around to complete the negative loop had failed. Inverted systems had yet to be invented.

The engine cut out as soon as the aircraft nosed beyond inverted; power was unavailable to assist in the push-up. Nor were pilots willing to accumulate the required energy in a dive, fearing structural failure. The machines of the day were thought to be too frail for the job and the maneuver was deemed to be simply impossible—just the kind of challenge to capture the imagination of Jimmy Doolittle.

By 1927, Doolittle had been flying a 435-hp Curtiss P1-B fighter for quite some time and had come to believe that here, at last, was a machine capable of withstanding the outside loop. Ever the professional test pilot, he undertook a long series of flights, pushing the airplane and himself a little further around the curve every time, carefully analyzing his experiences and scrutinizing the airframe after every flight. He kept his experiments very low key, letting only a handful of trusted pilots in on his plans. Characteristically, he saw no sense in stirring up the public before success was assured.

Finally, on May 25, 1927, over McCook Field in Dayton, Ohio, he climbed to 10,000 feet, pushed over into a vertical dive to the P1's redline, and kept right on pushing. The experiment had paid off. Soon he was upright once more in straight and level flight none the worse for wear. When assaulted by enthusiastic reporters at the airshow where Doolittle first performed the outside loop he reveled in telling them that he had just done it on a whim.

5
Aileron rolls

THE OBJECTIVE OF THE AILERON ROLL IS TO ROLL THE AIRCRAFT 360° around its longitudinal axis, maintaining a ballistic, corkscrew-like flight path along the longitudinal axis. It is one of the easiest and most enjoyable aerobatic maneuvers to fly. It is not considered a precision maneuver; hence, it is not an aerobatic competition maneuver.

The aileron roll is a useful initial training maneuver. Besides being very simple to perform, it gets the student used to unusual attitudes without exposure to any more stress (if done correctly) than the standard positive 1G experienced in upright straight and level flight.

Typical minimum entry speeds

Aerobatic trainer:	130 mph
Intermediate aircraft:	140 mph
Unlimited monoplane:	160 mph

UNDERSTANDING IT

The primary concern of the pilot rolling an aircraft 360° is altitude loss. When an aircraft is rolled from straight and level flight and the wing generates less lift as it is banked, the aircraft will lose altitude. The greater the bank, the less the lift generated by the wing. That is why even in a steep turn we have to add power to maintain altitude. In a vertical bank, the wing generates practically no lift at all. A fact little known to nonaerobatic pilots is that in a vertical bank the fuselage acts as an airfoil. Regardless of the inefficiency, it does generate some lift, which is nowhere near the lift generated by the wing in upright straight and level.

Things don't improve much in the inverted position. At a similar angle of attack, a wing with an asymmetrical profile (which most of them are) is very inefficient in the inverted position in comparison to right side up; therefore, the wing requires a much higher angle of attack than in an upright position to generate the lift required to maintain altitude (more about this under the chapters covering the slow roll and inverted flight).

Somehow the tendency to lose altitude in the roll has to be overcome, and in an aileron roll it is accomplished by pitching the aircraft up into a climb prior to rolling. The slight climb compensates for the altitude to be lost in the vertically banked phases of the flight. The pitch up also accomplishes something else. As the aircraft rolls it starts to descend, which means that the direction of the relative wind changes; however, in the inverted position, the high pitch angle is a mirror image of the pitch angle in the upright position. This means that—taking into account the change in the direction of the relative wind—we have managed to substantially increase the angle of attack, to a point where the inverted wing is capable of maintaining altitude.

As we again bank toward the vertical on our way to complete the roll, a little more altitude is lost (already compensated for by the brief preroll climb) and we more or less end up in upright flight at the same altitude where we started the maneuver.

On its ballistic, corkscrew-like flight path, the airframe continuously experiences normal acceleration, maintaining light, positive G all the way around the aileron roll.

FLYING IT

The aileron roll can be done from cruise speed in most aircraft, though it is customary to carry a few extra knots (check your manual). It is easy to perform and delightful if done well. For the inexperienced pilot, it is also easy to make mistakes that result in a substantial loss of altitude and an increase in airspeed, so it should be learned with the appropriate amount of respect. Let's use the left aileron roll as an example and see how it is flown.

1. Establish the proper entry speed and straight and level flight. Verify airspeed.

2. Pitch up the aircraft approximately 30° above the horizon and neutralize all controls. It is especially important to neutralize pitch (elevator) because if you don't, the aft elevator will pull the nose farther up and off the point to the left during the first quarter of the roll and will pull the nose down during the inverted segment.

3. Briskly apply full left aileron and hold it there, and momentarily apply left rudder to counter adverse yaw as you begin the roll.

4. Maintain full left aileron and simply hang on for the ride. Briskly but smoothly neutralize aileron as you roll level to complete the maneuver. Be especially careful to maintain neutral elevator throughout the roll. Resist any instinct to pull back stick (elevator) when inverted because this will cause the nose to drop and will result in an enormous amount of altitude loss airspeed increase.

COMMON ERRORS

Errors are most likely to occur due to the subconscious misplacement of the controls by the novice during the initial experience of truly unusual attitudes.

- **On pull-up the nose is not raised high enough.** Initially, a 30° pull-up looks like a lot more, and the student might not raise the nose up where it should be. The result is that in the inverted position the aircraft will be at a lower pitch attitude than required for a straight-and-level attitude; the nose will drop, airspeed will build impressively, and the aircraft will end up straight and level well below the altitude where it started. Learn to get the nose way up there where it belongs.

- **On pull-up the elevator is not neutralized prior to beginning the roll.** As described in the prior subsection, the up elevator will pull the nose off to the left, up during the first quarter initially and down in the inverted position, resulting in something more like a barrel roll. A variation of this error is to neutralize the elevator initially, but then inadvertently apply up elevator (aft stick) on the way around, with generally similar results.

- **Inadvertent reduction of aileron input on the way around.** The roll rate will slow considerably and might stop all together. A smooth and deliberate reapplication of aileron cures the problem immediately; however, if your aircraft is not equipped with an inverted system and you take out aileron in an inverted position, a disconcerting sputtering sound followed by silence will make it easy to commune with the whispering wind and will probably cure you of aileron input errors for the rest of your life. If the engine cuts out, promptly keep rolling and it will usually restart when you are upright again.

IF THINGS GO WRONG

Preventative steps are the key to avoiding unsafe situations from developing. Errors resulting in a great loss of altitude and speed are the most likely to compromise safety if there is insufficient altitude to correct them. Have plenty of extra altitude, way above FAA minimums, while you become completely comfortable with the maneuver.

Get that nose up where it belongs, even if belatedly; in the inverted position, this means applying slight forward stick as the nose drops and speed builds; upright it means slight aft stick. If for whatever reason you interrupted the roll, don't lose your composure upside down; simply resume rolling to upright straight and level. Avoid at all cost the temptation to pull through from inverted into level flight via the second half of a loop (a *split S*), instead of continuing the roll because your altitude loss and speed gain will be much higher, and potentially much more unsafe than in any botched aileron roll.

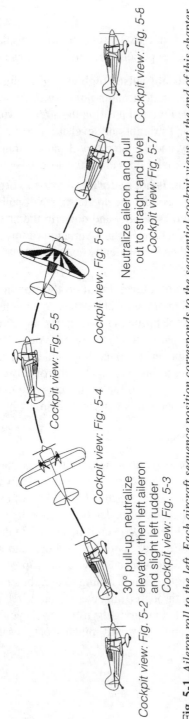

Cockpit view: Fig. 5-8

Cockpit view: Fig. 5-7

Neutralize aileron and pull
out to straight and level
Cockpit view: Fig. 5-7

Cockpit view: Fig. 5-6

Cockpit view: Fig. 5-5

Cockpit view: Fig. 5-4

Cockpit view: Fig. 5-3

30° pull-up, neutralize
elevator, then left aileron
and slight left rudder
Cockpit view: Fig. 5-3

Cockpit view: Fig. 5-2

Fig. 5-1. *Aileron roll to the left. Each aircraft sequence position corresponds to the sequential cockpit views at the end of this chapter.*

Fig. 5-2. *Straight and level at aileron roll entry speed.*

Fig. 5-3. *Thirty-degree pull-up. Neutralize controls, then roll with left aileron and slight left rudder.*

Fig. 5-4. *First quarter of roll complete.*

Fig. 5-5. *Inverted position. The pull-up prior to rolling establishes the inverted nose-high attitude, which allows the wing to generate sufficient lift in this position to maintain altitude.*

Fig. 5-6. *Third quarter of roll complete.*

Fig. 5-7. *Neutralize the controls and pull out.*

Fig. 5-8. *Exit the maneuver straight and level.*

6
Slow rolls

THE OBJECTIVE OF THE SLOW ROLL IS TO ROLL THE AIRCRAFT 360° AROUND its longitudinal axis while maintaining a level flight path. From the cockpit view, the goal is to roll the aircraft around the horizon while keeping the aircraft's nose on a specific point on the horizon. It differs from the aileron roll in that during that maneuver the aircraft's nose rolls in a ballistic path around a selected point on the horizon as the aircraft follows a corkscrew path around its longitudinal axis.

The slow roll is in part a negative G maneuver. While the aircraft is inverted it is subject to mild negative G.

The slow roll is a challenge. It is the first maneuver encountered by the aerobatic student that requires the simultaneous use of all three control surfaces: aileron, rudder, and elevator. It is an excellent training maneuver for learning and practicing the coordination skills required by typical aerobatic sequences. Unlike the aileron roll, the slow roll is an aerobatic competition maneuver.

It is worth noting that neither airspeed nor roll rate have anything to do with the name of the slow roll. Though the expression's origin is unclear, it was most likely inspired by the slow, graceful appearance of the maneuver when flown by the grand old barnstorming biplanes that popularized it during the 1930s.

Typical minimum entry speeds

Aerobatic trainer: 110 mph
Intermediate aircraft: 120 mph
Unlimited monoplane: 140 mph

UNDERSTANDING IT

To elaborate on the previous chapter's discussion of the aerodynamic effects of rolling, when the aircraft commences a roll from level flight initiated simply by full aileron control deflection in the desired direction of the roll, the lift generated by the wings diminishes as the angle of the roll is increased. Lift is no longer sufficient to counter gravity and the aircraft wants to descend, noticed in the cockpit when the aircraft's nose begins to drop.

As the aircraft goes inverted, the wing once again generates more lift than in banked flight, but not as much as it generates in upright level flight at an equivalent angle of attack (especially if the wing profile is asymmetrical); therefore, the aircraft will want to continue to descend. As lift is again diminished in the second half of the roll, the descent continues and stops only when the aircraft is once again in upright level flight; thus, the net result of a roll from level flight accomplished solely by the deflection of ailerons is a descending corkscrew maneuver ending in a marked loss of altitude.

This altitude loss has to be countered if the rolling maneuver is to be completed at the same altitude where it was begun. As we have seen in the section on aileron rolls, the simplest way to compensate is to pitch up the aircraft into a climb momentarily prior to rolling. This results in an angle of attack when the aircraft reaches the inverted position, which generates sufficient lift to maintain altitude; however, no attempt is made to compensate for the loss of lift during the banked phase of the roll, and the aircraft rolls around in a sloppy corkscrew pattern, eventually finishing up more or less where it started prior to pitching up at the commencement of the maneuver.

In the slow roll, in order to maintain a flight path that corresponds to the longitudinal axis in straight and level flight, the aircraft cannot be pulled up into a momentary climb prior to rolling. Instead, all three control surfaces must be used at various stages of the roll to counter the effects of diminished lift.

In the initial stage of the roll, as the aircraft approaches vertical bank, the lift generated by the wing practically ceases. Lift is now generated (poorly) by the fuselage acting as an airfoil. This lift is insufficient to maintain level flight. The nose would drop unless it is counteracted by top rudder. The upward deflected rudder, which is now in the horizontal position acting as the elevator, provides the angle of attack necessary for the fuselage to generate the lift required to hold the nose above the horizon.

As the aircraft rolls past the vertical, the elevator begins to regain its effectiveness as a method of pitch control. The aircraft is now approaching the inverted position, but is not generating sufficient lift to maintain altitude at the angle of attack it now has as a result of the position from which it commenced the roll. The nose continues to want to drop below the horizon. The angle of attack has to be increased. This is accomplished by pushing forward on the stick, which deflects the elevator in the direction above the horizon. This decreases the lift generated by the horizontal tail surface, and pushes the nose up above the horizon, increasing the angle of attack.

As the aircraft continues to roll and once again approaches the vertical, top rudder is again applied to keep the nose up. Past the vertical, forward stick (down elevator) is gradually brought aft to reach the neutral position as the aircraft nears straight-and-level flight. Aileron and rudder are neutralized at the last minute upon reaching upright level flight, and the slow roll is complete.

As is the case with all rolling maneuvers, at the commencement of the slow roll, a momentary application of rudder in the direction of the roll is required to overcome adverse yaw in most aircraft except some high-performance competition machines.

Airspeed in the slow roll is not that critical. Most aerobatic aircraft will do a slow roll comfortably from an airspeed approximately 10 percent above normal cruise speed (consult your aircraft manual). Generally a speed higher than recommended entry speed results in uncomfortably strong stick forces, while a slower entry speed can yield a sloppy maneuver.

G forces are not a major factor in the slow roll. While negative G is experienced in the inverted portion of the slow roll, G forces are light throughout the maneuver.

FLYING IT

Flying the slow roll requires a fair amount of concentration. Numerous simultaneous control inputs are required in a fairly brief time span, and the task is made more difficult by the fact that some aileron and rudder control input requirements are deliberately uncoordinated (for example, full left aileron and right rudder in the first stage of a left slow roll). At first it might be difficult to even recognize your mistakes regarding the finer points of the slow roll, let alone correct them, but you will regularly make it all the way around and in a few sessions you will soon start to get the hang of it.

A slow roll has seven basic stages; we will use a left roll as an example.

1. Pick a reference point on the horizon and position the aircraft on a heading pointing straight at it. Establish the entry speed, briefly transition to straight and level flight, and neutralize the controls. In training aircraft that have asymmetric wings, such as the Decathlon, pitch up slightly and neutralize controls after establishing the entry speed, but do not allow a climb to develop prior to commencing the roll. The pitch-up helps establish a high angle of attack approaching the inverted position and eases some of the force that will be required to push on the stick to keep the nose sufficiently above the horizon.

2. Commence a left roll with a smooth, full deflection of the ailerons, accompanied by a momentary application of left rudder (in lower performance training aircraft) to counter adverse yaw.

3. As you pass the 45° mark and approach vertical bank, gradually apply right (top) rudder to hold the nose in position relative to the horizon. The nose will want to drop as the amount of lift generated by the sharply banked wings diminishes, hence the need for top rudder.

4. As the aircraft rolls beyond the vertical to approximately 120°, the nose will again want to start dropping because the angle of attack is insufficient to generate the necessary lift for maintaining altitude. A firm but gradual forward motion of the stick is now required. As you roll inverted, you should have left aileron, right rudder to counter adverse yaw, and some forward stick pressure

to keep the inverted angle of attack high enough to prevent a loss of altitude. Now you can hold the applied control inputs until you roll past inverted flight.

5. When you roll past inverted and approach the 220° mark, the lift generated by the once again increasingly banking wing begins to diminish and the nose will again want to drop. A good dose of gradually applied left rudder will now keep the nose up on the horizon.

6. Continue to apply left (top) rudder as you roll past vertical on the way back to straight and level flight. Gradually ease off forward stick, returning it to the neutral position as you reach upright straight-and-level flight. Continue to maintain left rudder as required to keep the nose in position.

7. At the last instant before the aircraft is once again straight and level, briskly but smoothly neutralize the ailerons and rudder. If you did everything right, you should be at the same altitude and airspeed and on the same heading as you were when you started the slow roll.

COMMON ERRORS

The slow roll is one of the most demanding aerobatic maneuvers in terms of coordination and finesse, which is why it is such a good training maneuver for keeping your aerobatic skills sharp. Most slow roll errors are errors of coordination: not applying a particular control at the right time; overcontrolling or undercontrolling. The student needs a good instructor who can accurately point out these errors and their causes, and then it is just a matter of practice, practice, and more practice.

The most common errors are readily identifiable:

- **Insufficient top rudder, late application of top rudder.** This is the most common mistake. As the aircraft rolls on knife-edge, some top rudder is needed to maintain the fuselage's angle of attack sufficiently to maintain altitude. If insufficient rudder is applied, the nose will drop and the aircraft will assume a corkscrew-like flight path while loosing altitude.

- **Too much forward stick.** Some students are too enthusiastic in their efforts to apply forward stick to keep the nose up as the aircraft rolls inverted. Too much forward stick causes the aircraft to depart from its straightline flight path in a corkscrew fashion.

- **Increased roll rate near the end of the roll.** A common error is increasing the roll rate as the aircraft nears the end of the roll. There seems to be an urge to hurry the roll. While minor in nature and not compromising safety, it is the kind of error that can cost valuable points in competition.

IF THINGS GO WRONG

If you have sufficient altitude, little can go wrong during the slow roll to cause a safety problem. As with the aileron roll, the major concern is a great loss of altitude that results from common slow roll errors. Draggy, underpowered aircraft, such as most antique biplanes, are especially susceptible to altitude loss. Accidents caused by slow rolls gone wrong inevitably happen to unskilled or unqualified practitioners of aerobatics hotdogging at low level and literally running out of altitude.

Learn and practice at altitude, apply those rudders enthusiastically, and profit from your experience when the nose drops. Above all, if you want out of a slow roll, resist all temptation to pull through from the inverted position in a split-S; the altitude loss and airspeed gain would be tremendous. When a slow roll is started, the best way to get out of it is to keep right on rolling.

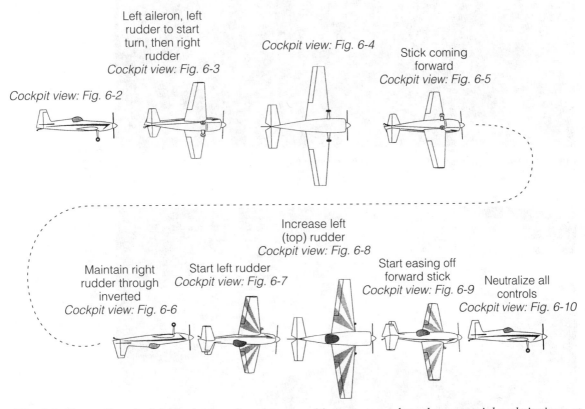

Fig. 6-1. *Slow roll to the left. Each aircraft sequence position corresponds to the sequential cockpit views at the end of this chapter.*

Fig. 6-2. *Establish straight and level at slow-roll entry speed.*

Fig. 6-3. *Left aileron, left rudder to start turn, then start right rudder.*

Fig. 6-4. *Stick coming forward, right rudder increasing.*

Fig. 6-5. *Stick farther forward to keep nose on horizon, maintain right rudder.*

Fig. 6-6. *Maintain forward stick and right rudder.*

Fig. 6-7. *Start left rudder.*

Fig. 6-8. *Increase left rudder.*

Fig. 6-9. *Start easing off forward stick.*

Fig. 6-10. *Neutralize all controls.*

Gerhard Fieseler and staying upside down

When the late Jimmy Doolittle was pushing his P-1B's nose over farther and farther in his 1927 attempt to conquer the outside loop, he was not alone in his quest. On the other side of the Atlantic, one of Europe's star aerobatic pilots, the German Gerhard Fieseler was pursuing the same goal by different means. His approach, and the time it took to develop it, would cost him the honor of being the first to do the outside loop, but would open a whole new world to the sport of aerobatics.

A big obstacle to doing the outside loop was the lack of inverted fuel and oil systems. As the aircraft nosed past the vertical, the engine would sputter and stop, denying its power to the airframe attempting to claw its way up the back half of the outside loop. If he could keep the engine running inverted, Fieseler reasoned, he might just have the extra push it would take to make it all the way around. So he set out to design and develop an inverted fuel and oil system for his Schwalbe biplane. The task proved to be relatively easy, except for the carburetor, a sensitive, precision component that gave plenty of trouble in those days even when it was right side up.

The solution proved to be, after interminable tinkering, an early form of fuel injection. Fieseler was soon merrily flying upside down under power for prolonged periods and began to cautiously work up to conquering the outside loop. He finally did one in June 1927—Doolittle beat him by less than a month.

But Fieseler, too, got his reward when later that summer he was the first to dazzle the airshow crowds with prolonged inverted maneuvers under power. The use of inverted systems spread rapidly in the aerobatic community and inverted flight as we know it today was here to stay.

Another contribution by Fieseler to competitive aerobatics was the rolling circle maneuver, which he first flew in 1929 and which remains one of the most challenging advanced maneuvers to this day. In 1934, he won the first world aerobatic championships in Paris, France, and has remained active in the world of aerobatics into the 1990s.

In nonaerobatic circles, Fieseler is best known for designing the remarkable STOL military liaison aircraft widely used by the Germans during World War II, the Fieseler Storch.

7
Inverted flight

INVERTED FLIGHT IS OPERATING THE AIRCRAFT UPSIDE DOWN FOR A sustained period of time. It is a negative-G maneuver, and can therefore be performed only in aircraft equipped with inverted fuel and oil systems. In terms of our definition, the inverted portions of such maneuvers as rolls and loops do not constitute true inverted flight. While in inverted flight, the student is expected to be able to fly straight and level, climb and descend, and perform turns. It is an essential aerobatic skill, to be acquired by all aspiring aerobatic pilots.

While aerobatic sequences consisting entirely of positive G maneuvers can be flown in aerobatic aircraft that are not equipped with inverted systems, a botched positive-G maneuver in any aircraft might result in negative-G inverted flight, a compelling reason for all aerobatic pilots to learn to fly inverted. Inverted flight is an important building block of competition aerobatic sequences.

Typical inverted cruise speeds

Aerobatic trainer:	125 mph
Intermediate aircraft:	145 mph
Unlimited monoplane:	170 mph

UNDERSTANDING IT

The most obvious question regarding inverted flight is how a wing, especially one that is asymmetric, is able to generate lift in the inverted position. A very basic review of how a wing generates lift is in order. First, consider the airflow over an asymmetric wing. The greater curvature of the wing's top surface provides a longer distance for the air particles to travel in the same time that particles travel along the wing's underside. In upright straight-and-level flight, the air particles flowing over the top surface speed up in comparison to the particles flowing along the underside of the wing. As per Bernoulli's principle, the faster airflow results in lower pressure along the top surface in comparison to the bottom surface, and the wing is suspended in the airflow by the combined effect of the two pressure zones.

Now we have to once again talk of angle of attack. If we increase the asymmetric wing's angle of attack, we displace on the leading edge of the wing the point where the airflow separates to traverse the wing's upper and lower surfaces. By increasing the angle of attack, we have further increased the length of the top surface and decreased the length of the bottom surface along which the air particles have to travel. The airflow along the upper surface speeds up further relative to the airflow along the lower surface, and the result is increased lift.

So, what about a symmetric wing, the wing found on most high-performance aerobatic aircraft? If the upper and lower surface curvatures are the same, the pressure zones on the two sides should be equal and the wing should never be capable of ascending. Yet it is, and the answer is angle of attack. A 2–3° angle of attack will sufficiently move the point on the leading edge where the airflow separates to create a difference between the length of the wing top and bottom surface over which the air flows, resulting in an increase in lift.

Now consider the asymmetric wing inverted. The task is to increase the distance over which the air particles have to flow along the top surface to exceed the distance traveled by the airflow along the bottom surface. This is done by increasing the angle of attack. The increase has to be substantial, which is why aircraft with asymmetric wing profiles can maintain inverted level flight only in a noticeably nose-high attitude.

The principle is the same for the symmetric wing, but the angle of attack required is small and equivalent to that required right side up. The symmetric wing performs equally well either side up, which is the reason why it is found on high-performance aerobatic aircraft.

FLYING IT

Inverted flight can be entered in several ways. You can roll inverted, you can pull up inverted through the first half of an inside loop, or you can push over into inverted flight through the first half of an outside loop. Rolling inverted is the easiest and least stressful way to enter inverted flight, so it is taught here. Once you feel comfortable flying upside down and have learned to do inside loops, you will routinely pull up through a half loop into inverted flight. Pushing over into inverted flight through an outside half loop is an advanced maneuver and should be left to the advanced stage of your aerobatic training.

Another important aspect of inverted flight is recovering from it. You can roll upright, which is the preferred recovery technique, or you can pull through the second half of the loop, which is the instinctive reaction but is emphatically not recommended as a standard recovery technique because of the rapid airspeed buildup, high acceleration load (Gs), and loss of altitude.

Let's learn the maneuver, how to roll into it, and how to recover from it.

1. Rolling inverted. We have already learned how to do rolls. There is nothing more to rolling inverted than simply stopping a roll by neutralizing aileron and rudder control when the aircraft reaches the inverted position.

    As we emphasized in the sections covering rolls and in the presentation above, you need a high angle of attack (especially with an asymmetrical wing) in the inverted position to maintain level flight (i.e., to stop the nose from dropping). There are two options in establishing this angle of attack. You can roll inverted and push the nose up, which is an uncomfortable negative G procedure, or you can pitch up before you start the roll and as you reach the inverted position you will be already at the appropriate angle of attack. This latter method is more comfortable and highly recommended.

    Thus, the simplest entry is a half aileron roll. Pitch up, neutralize controls, apply full aileron and momentary rudder in the desired direction of the roll, and neutralize aileron when you reach the inverted position.

    You will also find it easy to enter inverted flight through a half slow roll. As we have learned, the slow roll requires pushing the nose up as the aircraft rolls past the vertical, but forward stick is applied gradually, thus minimizing the effect of subjecting our body to negative G.

2. Inverted straight and level. When you are established in the inverted position, the first order of business is to get used to straight and level. Glance at the airspeed indicator to confirm stable and appropriate airspeed and look straight ahead at the horizon as you did when you first learned straight and level right side up. The negative G sensation might feel a bit weird (you will think you are hanging from your straps far more than you really are) and in a training aircraft the nose will be way above the horizon. The switched places of earth and sky will most likely seem more normal than you had expected.

    Practice maintaining straight and level flight. When you are used to the general view, glance more frequently at the airspeed indicator and the vertical speed indicator for backup confirmation of your attitude. Gently experiment with shallow climbs and descents. Note the strong stick forces necessary to keep the nose up, and the eagerness of the aircraft to descend. Should you retrim in inverted flight? Generally, you shouldn't. You will be spending only a few minutes inverted, and if you retrim, you will have to retrim again for the subsequent maneuvers of your session.

3. Inverted turns. Doing turns in inverted flight is confusing at first. You have to bank in the direction you wish to turn as you do in upright turns, which requires aileron deflection in the same direction; inverted or upright, your aileron input is always in the direction of the turn. To turn left, you need to apply left aileron, upright or inverted. So far, so good. Now you have to overcome the adverse yaw created by the wing outside the turn (the up wing). Upright, you would apply left rudder, the rudder on the inside of the turn. To

achieve the same effect upside down, you also have to apply the rudder on the inside of the turn; however, upside down, this rudder is the *right rudder*, which in mirror image is where the left rudder was upright.

To summarize: In an inverted turn you have to apply aileron in the direction of the turn and rudder opposite the direction of the turn. This is truly easier to understand and learn by doing than by reading about it. With some trial and error you will soon get the hang of it.

4. Recovering from inverted flight. The best way to recover from inverted flight is to roll right side up. You should practice this mode of recovery until it becomes instinctive. In emergencies, it is much safer than pulling down into the second half of a loop. The roll puts a negligible acceleration load on the aircraft, airspeed does not increase, and little altitude is lost even if the maneuver is done in a sloppy fashion.

Pulling down into a half loop will, on the other hand, lead to a rapid speed buildup, it will require pulling a substantial acceleration load to control the speed buildup, and it will result in a substantial loss of altitude even if done correctly. You will learn to incorporate into aerobatic sequences a planned downward half loop from inverted flight, but as a standard or emergency recovery technique from inverted flight, stick to rolling upright.

When you are ready to recover from inverted flight, smoothly apply full aileron in the desired direction of the roll accompanied by momentary opposite rudder to correct adverse yaw and wait for the horizon to turn right side up. Gradually neutralize forward stick as you roll around. If you want to be fancy, you may turn the recovery into the second half of a slow roll by adding top rudder as you approach the first 45° of roll and maintaining it until you are about 45° from the upright position.

COMMON ERRORS

Only one common error is committed by novices learning inverted flight: the failure to maintain a sufficiently nose-high attitude to maintain inverted straight and level. In training aircraft, the required attitude is extreme, and the control pressures are high. The discomfort of negative G does not help matters. To learn inverted straight and level, revert to what you did on your first instructional flight when you were learning to fly. Concentrate on the horizon and keep it where it belongs. Ignore the fact that the space above the horizon is now ground and the space below, sky. Remind yourself that whatever you feel is natural, however unusual it might initially seem to you.

IF THINGS GO WRONG

One reason pilots are initially reluctant to get the nose way up in inverted flight is their fear of an inverted stall, possibly developing into an inverted spin. From straight and level, an inverted stall might be encountered, but is about as benign as an upright stall. The nose will drop and the aircraft will soon be flying again. Remember, in inverted flight you need to apply *aft* stick (elevator) to get the nose down. So to recover from a stall, make sure the rudders are neutral and apply aft stick. The forward stick forces are so high that you will encounter plenty of prestall buffet in most training aircraft before experiencing an inadvertent inverted stall.

An inadvertent inverted spin is possible from inverted straight and level flight with the gross misapplication of controls (extreme uncoordination), but the chances of it developing are quite remote in most low-performance training aircraft. In a badly mishandled turn, the chances are slightly higher. The standard recovery technique from the inadvertent inverted spin is:

1. Power off.
2. Rudder opposite the *yaw* to stop the rotation.
3. Aft stick to unstall the wing.
4. When the spin stops, neutralize the controls and pull out of the dive.

The Beggs/Müller emergency recovery technique also works in most aircraft:

1. Power off.
2. Let go of the stick.
3. Rudder opposite the *yaw* (step on the rudder that offers the greater resistance).
4. When the spin stops, neutralize the controls and pull out of the dive.

It is imperative that you get comprehensive inverted stall/spin instruction from a qualified aerobatic instructor prior to solo inverted flight.

Things can go very wrong if for whatever reason you decide to abandon inverted flight and decide to pull through in a split-S instead of rolling upright. The altitude loss and airspeed gain in a split-S can be tremendous and a hazard to safety. Always roll upright to recover from inverted flight.

A WORD ABOUT INVERTED SYSTEMS

An inverted system is essential for prolonged negative-G flight, such as inverted flight, and an aircraft has to be specifically approved for inverted flight to be deliberately flown inverted, so a few words about inverted systems are worthwhile.

The direction of the flow of fluids circulating in an aircraft is affected by the force of gravity. In inverted flight, the fluids will flow in a different direction relative to the aircraft than in upright flight. Unless the systems using these fluids are specifically designed to take this difference into account in inverted flight, the flow of fluids will be interrupted and the systems will cease to function.

Fuel and oil are a problem in inverted flight on every aircraft, and on some sophisticated (mostly military aircraft) hydraulic fluid can also cause a problem. If there is no inverted system, when an aircraft is turned upside down, fuel and oil will cease to flow through the appropriate parts of the engine, which will stop and could even be damaged.

Fluid flow can be maintained many ways when flying inverted, but inverted systems generally fall into two broad categories: *time-limited* systems and *continuous* systems. Time-limited systems are based upon an inverted auxiliary tank. Fluid collects in the inverted tank during upright flight, and the tank is so positioned that when the aircraft is turned upside down it is able to supply fluid to the appropriate systems until it runs dry. These systems are simple and usually provide fluid for not much more than 5–10 minutes at a time, which is more than it might seem considering that the average length of an unlimited competition sequence is only about 8 minutes.

The more complex inverted systems innovatively combine a network of pumps and plumbing to provide a continuous supply of fluid equal to the supply available right side up. You should be intimately familiar with your aircraft's inverted capabilities and system. Consult your aircraft manual.

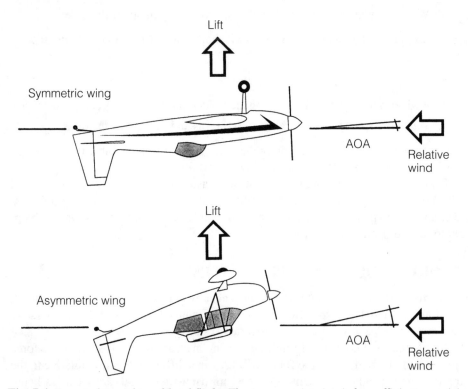

Fig. 7-1. *Inverted straight and level flight. The asymmetric wing is less efficient, requiring a higher angle of attack (AOA) to maintain straight and level.*

Fig. 7-2. *Upright straight and level cockpit view from an aircraft with a symmetric wing.*

Fig. 7-3. *Upright straight and level cockpit view from an aircraft with an asymmetric wing.*

Fig. 7-4. *Inverted straight and level cockpit view from an aircraft with a symmetric wing. Note nose-high atltitude.*

Fig. 7-5. *Inverted straight and level cockpit view from an aircraft with an asymmetric wing.*

How the Cuban eight got its name

Len Povey was a colorful American barnstormer of the 1930s whose pyrotechnic night aerobatic routine caught the attention of Cuban military officials. They were looking for a pilot to start the Cuban Air Force and Povey got the job.

In his new role, he flew a Cuban Air Force Curtiss Hawk to Miami to participate in the 1936 All American Air Race Meetings. While performing the freestyle aerobatic routine for the Freddie Lund Trophy, he intended to do a triple snap roll off the top of a loop, but realized he was much too fast. So he rode the loop over the top and coming down the back, on a whim, he decided to do a half roll and immediately pull up into an opposing loop followed by another half roll on the back end on the way down. Asked on the ground what the maneuver he just did was supposed to be, the quick-thinking Povey casually remarked that it was a "Cuban eight."

8
Loops

THE LOOP IS A 360° CIRCLE FLOWN BY THE AIRCRAFT IN THE VERTICAL plane. There are two variants of the loop: the inside loop and the outside loop. During the inside loop, the top surface of the aircraft is on the inside of the circle flown by the aircraft; during the outside loop, the top surface of the aircraft is on the outside of the circle flown. The inside loop is a positive-G maneuver; the outside loop is a negative-G maneuver. This chapter covers only the inside loop (in no wind), which is a basic aerobatic maneuver. The outside loop is considered an advanced aerobatic maneuver due to the high negative G to which it subjects the aircraft and the pilot. The loop is an important element in competition aerobatics.

Typical minimum entry speeds
Aerobatic trainer: 140 mph
Intermediate aircraft: 160 mph
Unlimited monoplane: 170 mph

UNDERSTANDING IT

The main challenge faced by an aircraft in performing a loop is to sufficiently overcome the force of gravity to arch over the top of the loop; thus, the loop is an exercise in energy management. The aircraft must build sufficient speed to store the energy required to complete the loop. It must then expend this energy at a high enough rate to fly over the top of the loop.

The minimum acceleration rate required to complete a loop from the entry speeds stated in the flight manuals is around 3.5G. The maneuver may be flown from the given entry speed at a higher acceleration rate, in which case the circumference of the circle will be proportionally smaller.

If the loop is attempted at an acceleration rate below the minimum required, the aircraft will run out of airspeed prior to reaching the top of the maneuver, the wing will exceed the critical angle of attack, and it will stall. If the stall occurs beyond the vertical plane, the aircraft's nose will most likely pitch down over the top and the aircraft will complete a narrow, elliptical figure that will resemble a highly squashed loop, but will be an incomplete maneuver. Depending upon aircraft type, a wing might also drop, resulting in some interesting views of real estate.

Entry speeds for a loop might be higher than the minimum entry speed (competition and display pilots often enter loops at higher than minimum entry speed). From a higher entry speed, the standard 3.5G pull-up will yield a loop with a greater circumference. If a tighter circumference is desired from a higher entry speed, a proportionally higher acceleration rate is required on the pull-up into the loop.

When the aircraft is pulled up into the loop, acceleration increases and airspeed dissipates. Going over the top of the loop, the airspeed is at its lowest value for the maneuver. As the aircraft starts back down on the second half of the loop, airspeed increases once again, and could easily build to an excessive value if not checked.

Remember, a buildup of airspeed is a buildup of energy, and the way to dissipate it or to keep it from increasing is to load it down with acceleration (Gs). This is accomplished by the pull-out from the loop, when, by pulling the same amount of acceleration as on the way up you will finish the maneuver at approximately the same airspeed and altitude where you started it.

Another important aerodynamic consideration in understanding the ability of an aircraft to loop is relative wind and angle of attack. The reason the aircraft can continue to fly throughout the loop—in spite of its extreme attitude relative to the ground—is that with respect to the relative wind, it continues to maintain an angle of attack below the critical angle of attack throughout the maneuver. For the appropriate angle of attack to be maintained, the relative wind must be able to keep pace with the rate of the aircraft's pitch change throughout the maneuver. It is, therefore, important to avoid abrupt control movements and to apply all controls smoothly and consistently during the loop.

FLYING IT

Flying the loop is a delicate balancing act of airspeed and acceleration control. It is the first maneuver during which the student pilot is exposed to high G loads and approaches certain performance limits of the aircraft, especially if the execution is not near perfect. With a little practice it is easy to fly, and once they feel comfortable with it, most aerobatic pilots find it great fun and count it among their favorite maneuvers. Let's see how it is flown.

1. Loops must be performed along a longitudinal reference line, such as a road, railroad, or power line, so that when you arch past inverted flight and start down the back of the maneuver, you have a directional reference; thus, the

first order of business is to find a suitable reference line and line up over it. Be aware that the reference line you will be using will be the portion of the line directly beneath you, not the section you can see over the nose in straight and level flight.

2. Increase airspeed to the aircraft's loop entry speed, establish momentary straight and level flight, and neutralize the controls. *Verify airspeed*. Take special care to ensure that the wings are level. You will actually have to hold slightly forward stick to maintain level flight at the higher speed because you didn't—and shouldn't—retrim for the entry speed. By *neutralizing the stick*, we mean holding it motionless in the position it takes to maintain level flight at the selected loop entry speed.

3. Commence a steady 3.5G–4G pull-up by applying smooth back pressure on the stick. While you are learning the maneuver, a little extra acceleration to 4G is a good idea to give you more room to correct any errors. Learn to do the loop without looking at the G meter. After a few tries your body will tell you the amount of G you are pulling. The trick to doing a pretty loop is to apply steadily increasing back pressure on the stick on the way up, so that you reach the aftmost stick position appropriate for the maneuver (*not* full aft) only at the key point 30° beyond vertical (step 5 explains the key point). Steadily increasing back pressure is required to keep the radius of the loop constant because as the airspeed slows on the way up, the elevator becomes less effective.

4. As soon as the horizon disappears under the cowling, look out at the left wingtip, which now becomes your reference point. The wingtip indicates three things:

   ~ The angle of the wing chord relative to the horizon indicates your position on the circumference of the loop.
   ~ The extent to which the chord is on, below, or above the horizon indicates if your flight path is along the vertical plane or if you are yawing left or right. In a Decathlon, for example, the wing chord should bisect the horizon evenly. You can correct a yaw error by appropriately repositioning the wing chord with rudder and aileron.
   ~ The rate at which the wing chord line rotates around the horizon is an excellent indication of the loop's progress. A rate that is too slow indicates that you are not pulling sufficient acceleration to fly a nice, round loop. Continue to gradually apply back stick.

5. When the aircraft reaches 120° in the loop (30° beyond vertical) as indicated by the wingtip, you reach the *key point* in the loop. This point will be an important reference point for doing Immelmanns and Cuban eights.

   At the key point, look forward and above through the greenhouse roof or canopy to watch for the horizon to appear, and ease off slightly on the stick. Do not make the mistake of continuing to push on the stick—the loop will then stop and you will be flying inverted)—just ease off and hold it there. Easing off the elevator is necessary because as the G forces decrease near the top of the loop with the decrease in climb, it takes less elevator to give the same pitch effect. If you did not ease off the elevator, the aircraft's nose would pitch down the back of the loop prematurely and your loop would have an ugly oval shape.

Having eased off the stick, continue to look out the top, watching for the horizon to appear, and let the aircraft coast gracefully over the top of the loop.

6. As the aircraft goes over the top, you will first see the horizon and then the reference line below on the ground, which will give you a good indication of how accurately the aircraft is aligned in the maneuver.

7. As the aircraft approaches the 210° point in the loop (30° beyond the top), it begins to accelerate at a rapid rate. Watch the reference line (make corrections with the rudder, if necessary) and begin to apply steady back pressure on the stick to prevent the airspeed from building up and to pull out of the loop. Smoothly pull an acceleration rate equivalent to the rate you pulled at the commencement of the maneuver, and you should finish the loop straight and level at the same altitude and airspeed at which you started the maneuver.

A technique that some pilots find useful in performing aerobatic maneuvers is talking themselves through them. Everyone can work out a suitable personal monologue, but this "key list" script might be helpful:

- Line up over reference line

- Establish entry speed, level flight, *verify speed*

- Pull! Look left at wingtip as horizon disappears

- Keep pulling, vertical

- *Key point*, ease off stick, look ahead, up

- Horizon, float over the top

- Coming down, start pulling stick, ground alignment, straight and level, glance at airspeed

COMMON ERRORS

A pilot can commit several common errors while doing a loop.

- **Failure to continue gradually increasing back pressure on the stick on the way up as the aircraft slows and the elevator becomes less effective.** This is the most common pilot error. The result is an elliptical shaped loop and a low score in competition. Pilots who do not receive proper aerobatic instruction often merrily loop their way across the skies in this fashion, blissfully unaware of the eyesore loops that they are flying.

- **Failure to ease off aft stick at the key point.** If the pilot does not ease off aft stick at the top of the loop, the nose will swing down too early at too fast a rate and the figure will be egg shaped. An excessive increase in airspeed might develop and excess altitude might be lost.

- **Failure to establish or maintain horizontal wing alignment.** This is a fairly common error. As a result, the aircraft does not stay aligned in the vertical plane relative to the ground and will be off heading on recovery. A good ground reference line and a willingness to use it can cure this error quickly.

- **Failure to pull up sufficiently aggressively, resulting in a loss of too much airspeed at or before the key point.** This error also results in a lopsided, elliptical loop when the aircraft stalls out inverted and the nose drops sharply down into a dive. Some pilots don't even realize they actually stalled, as the nose swings back downward. They think they didn't ease off the stick sufficiently at the top. Quality instruction clarifies such misconceptions. If the wings are not horizontal in the loop, a wing might also drop under this scenario.

IF THINGS GO WRONG

On the way up, it takes a lot of work to get the aircraft into a position to compromise safety. The worst event that can happen is stalling out inverted, with possibly a wing dropping. Neutralizing the controls will soon have the nose pointed at some recognizable attitude, usually the ground in a dive, from which a normal recovery can be made.

On the way down, excessive G buildup, airspeed increase, and altitude loss are potential hazards. *Reduce power well before Gs and airspeed become a problem!* Don't exceed G limits with an aggressive pull out, avoid exceeding redline (V_{ne}), but don't allow your fear of redline cause you to pull excessive Gs, and always have a very generous altitude reserve, so that running out of altitude is never even a potential problem.

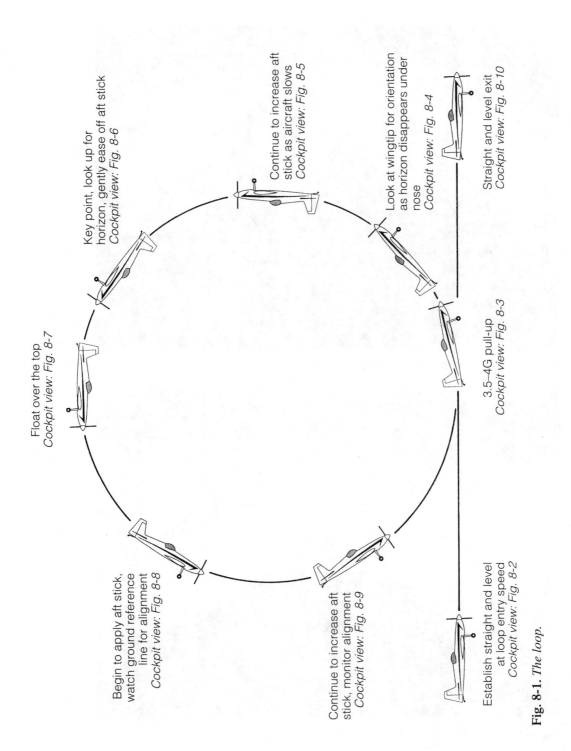

Fig. 8-1. *The loop.*

Continue to increase aft
stick as aircraft slows
Cockpit view: Fig. 8-5

Look at wingtip for orientation
as horizon disappears under
nose
Cockpit view: Fig. 8-4

Straight and level exit
Cockpit view: Fig. 8-10

Key point, look up for
horizon, gently ease off aft stick
Cockpit view: Fig. 8-6

3.5–4G pull-up
Cockpit view: Fig. 8-3

Float over the top
Cockpit view: Fig. 8-7

Establish straight and level
at loop entry speed
Cockpit view: Fig. 8-2

Begin to apply aft stick,
watch ground reference
line for alignment
Cockpit view: Fig. 8-8

Continue to increase aft
stick, monitor alignment
Cockpit view: Fig. 8-9

Fig. 8-2. *Establish straight and level at loop entry speed.*

Fig. 8-3. *3.5–4G pull-up.*

Fig. 8-4. *Look left at wing as horizon disappears underneath the nose.*

Fig. 8-5. *Continue increasing aft stick as aircraft slows.*

Fig. 8-6. *Key point. Ease off aft stick, transfer vision forward to look for horizon.*

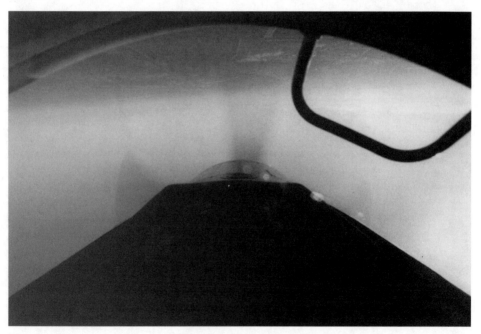

Fig. 8-7. *Float over the top, use horizon for wing alignment.*

Fig. 8-8. *Begin to apply aft stick, use ground reference lines for alignment. (Note that a grid of thin lines works as well as one solid reference line.)*

Fig. 8-9. *Continue to increase aft stick.*

Fig. 8-10. *Pull out into straight and level.*

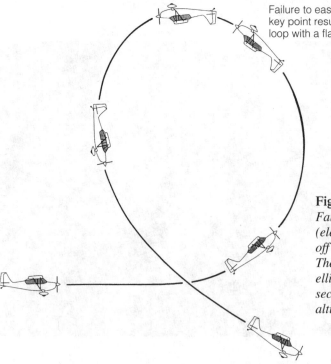

Failure to ease off aft stick at
key point results in elliptical
loop with a flat second quarter

Fig. 8-11. *Common loop error.
Failure to ease off aft stick
(elevator) at the key point cuts
off the second half of the loop.
The result is an imperfect,
elliptical loop with a flat
second quarter and possible
altitude loss on the pull out.*

Mad Max on your tail

Among the few pilots who have an aerobatic maneuver named after them, Max Immelmann, the World War I German fighter ace is perhaps the best known. It is ironic that he never flew the maneuver bearing his name. His Fokker monoplane fighter's frail structure would have never withstood a pull-up into a half loop followed by a half roll to upright straight and level flight, the maneuver now commonly called the Immelmann. Max did, however, perfect quick reversal techniques and his story is a good illustration of how the necessities of aerial warfare led to refinements of aerobatic technique.

Immelmann was among the first fighter pilots to discover that the key to success lay not in heroic, drawn out duels to the finish, but in the element of surprise. If your opponent notes your presence only when your ordnance hits the prey's airframe, you'll live to fight another day. The advantage of diving on his opponent from behind was obvious to Immelmann. The key to an undetected getaway appeared to be in a quick flight path reversal following the firing pass, to avoid overshooting the target and running the risk of turning into the prey. But how could the reversal be accomplished?

Speed had to be dissipated to permit a small turning radius and a turn away from the target was essential to avoid an overshoot. Immelmann realized that the most effective way to accomplish these objectives was to pull the nose up sharply and to begin turning at the same time. In developing his technique, Immelmann actually perfected two maneuvers common today, the chandelle and the wingover. The chandelle, a 180° climbing turn, was useful when his objective was to gain back as much height as possible following his diving attack, but it left the airplane precariously slow and vulnerable to counterattack. If a speedier retreat was of the essence, he flew a wingover, pulling up the nose, applying rudder, and gaining speed for the getaway as the nose swung around and pointed at the ground.

Immelmann's tactics worked especially well against the British BE 2 and the French Voisin. Both types were widely used as observation aircraft and primitive bombers in the early days of World War I. Both had pusher engines and a gunner/observer in front of the pilot in the nose, leaving them wide open to attack from behind. Immelmann racked up 15 victories and developed quite a reputation for stealth, but in the end fought one battle too many.

As his story was retold over the years, the chandelle and the wingover became a roll of the top of a half loop, or, an Immelmann.

9
Half loops

DURING THE HALF LOOP THE AIRCRAFT IS FIRST PULLED UP INTO A LOOP. Upon completing the first half of the loop, it is transitioned into inverted flight briefly. To recover, it is then half slow rolled into upright straight and level flight.

The half loop is the first combination maneuver encountered by the basic aerobatic student. It combines elements of the loop, inverted flight, and the slow roll. These three maneuvers must be fully understood and flown comfortably before the half loop is attempted.

The half loop is a very important aerobatic maneuver because it is frequently used as a building block in aerobatic sequences. It teaches the pilot slow-speed transition and slow-speed inverted flight, which are fundamental elements of capping off a vertical maneuver.

The half loop fulfills another important role. It is the most commonly used technique to set up the aircraft for inverted spinning in training and competition. It is preferred to rolling inverted to set up the spin. From a half loop the aircraft achieves inverted flight near stall speed. Instead of being rolled upright to complete the maneuver, the aircraft can immediately be put in an inverted spin. From a half roll, a time-consuming, inverted, straight and level segment must first be flown to slow down the aircraft to the proper spin entry speed.

Typical minimum entry speeds

Aerobatic trainer:	160 mph
Intermediate aircraft:	160 mph
Unlimited monoplane:	180 mph

UNDERSTANDING IT

There is little to discuss regarding the theory of the half loop because its elements are already familiar from the sections covering the loop, inverted flight, and the slow roll. During the pull-up into the half loop, the aircraft must build slightly more energy as it does during the full loop. The extra energy is necessary to make it easier to transition into inverted flight at the top of the loop; therefore, the half loop should be entered approximately 10 mph faster than the full loop, depending upon aircraft type (a draggy old Stearman will require a lot more speed) and the pull-up should be in the order of 3.5G–4.0G in a training aircraft (higher in a high-performance monoplane, for the sake of a tight maneuver).

The transition to inverted flight must be precise. Decisive forward stick at the right moment to stop the aircraft's nose at the inverted straight and level attitude above the horizon is a key element in properly performing the half loop. The aircraft is slow and close to the critical angle of attack (especially a low-performance training aircraft); therefore, to avoid a stall, great care must be taken not to stop or subsequently raise the nose too far above the horizon. While the forward movement of the stick should be decisive, it must not be shoved forward so rapidly as to cause an accelerated stall.

When the aircraft is in inverted flight, it is allowed to build up airspeed to inverted cruise speed. It is then slow rolled right side up, exactly as the second half of a slow roll is performed, which is explained elsewhere in this book.

FLYING IT

Preparations for the half loop are no different than they are for the full loop. The most important thing is mental preparation, given the combined elements of the loop, inverted flight, and the roll, and the need to anticipate each element. Let's see how the half loop is flown.

1. Establish the proper entry speed in a shallow dive and go into level flight momentarily, neutralizing controls. *Verify airspeed.* As in the loop, the stick will have to be slightly forward of neutral to maintain level flight because the aircraft is flying at a faster speed than for which it is trimmed.

2. Pull up into the loop. Be sure to gradually keep pulling the stick farther and farther aft because as the aircraft slows, the elevator becomes less effective, and increasing back pressure is required to keep the loop's radius constant.

3. As the horizon disappears under the cowling, transfer your vision to the left wingtip, and, as in a full loop, watch for the wing chord to be 120° relative to the horizon (30° past the vertical). This is the key point.

4. At the key point, transfer your vision above your head to watch for the horizon, ease off the stick, and, as the nose approaches (descends toward) the proper straight-and-level inverted attitude, push the stick forward deliberately to stop the nose at this inverted attitude.

5. Allow the aircraft to pick up speed in inverted straight and level flight. Glance at the ASI and VSI to confirm straight and level cruise.

6. Upon reaching cruise speed, slow roll upright as you have already learned to do. Rolling to the left: full left aileron, simultaneously momentary right rudder followed by a good dose of left rudder, gradual aft stick toward neutral with about 30° to go, neutral aileron and rudder at the last moment, and the half loop is complete. Rolling to the right, the control inputs are in the opposite direction. Over time you should be able to roll equally comfortably in both directions.

Half loop key list

- Line up (over reference line)
- Establish entry speed, level flight, *verify speed*
- Pull! Look left (at wingtip as horizon disappears)
- Keep pulling, vertical
- *Key point*, ease off stick, look ahead, up
- Inverted straight and level attitude, stick briskly forward to straight and level position
- Establish straight and level, maintain . . .
- Roll upright

COMMON ERRORS

Being a composite maneuver combining elements of the loop and inverted flight, the half loop's common errors are in large measure similar to those often seen in the performance of the first half of the full loop and during inverted flight (refer to the loop and inverted flight chapters for a refresher).

One common error peculiar to the half loop is a failure to precisely judge when to transition to inverted flight at the top of the loop. The trick is to be able to accurately judge the inverted straight and level flight attitude and swiftly and precisely apply forward stick in just the right amount a split second before the correct attitude is reached. If the transition is too soon, the aircraft might end up nose high, shuddering on the edge of an inverted stall. If the transition is too late and then the error is corrected too aggressively, an inverted accelerated stall might result. If the transition is delayed and the correction is applied gradually, the only consequence will be bruised pride for having poorly flown the maneuver. A discreet windshield mark indicating the proper position of the horizon in inverted straight and level flight can be a big help in flying the half loop properly.

Another common error is pushing the stick too far forward, though correctly judging the point where the inverted straight and level attitude is reached. The nose is pushed too high and the aircraft stalls.

IF THINGS GO WRONG

Safety might be compromised during the half loop just as it might be during its component maneuvers, the first half of the loop and inverted flight. The most common culprits are inverted stalls, accelerated stalls, and perhaps even a rare inadvertent spin if the controls are especially mishandled. By the half loop stage, you should be able to cope with each of them based upon your experience gained during previous maneuvers.

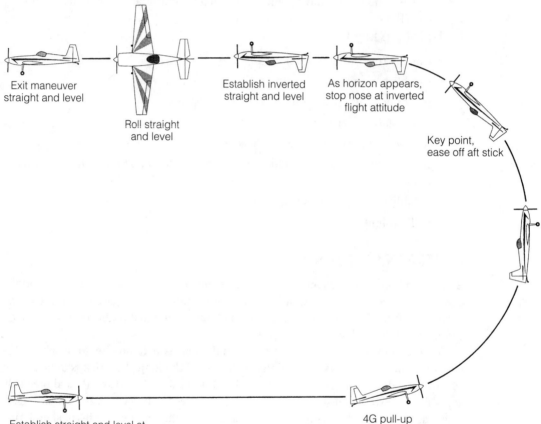

Exit maneuver
straight and level

Roll straight
and level

Establish inverted
straight and level

As horizon appears,
stop nose at inverted
flight attitude

Key point,
ease off aft stick

4G pull-up

Establish straight and level at
half-loop entry speed

Fig. 9-1. *The half loop. The roll at the top is to the right.*

10
The Immelmann

THE IMMELMANN IS A PULL-UP INTO A HALF LOOP, FOLLOWED BY AN IMmediate slow roll upright from the inverted position, the moment the aircraft reaches the top of the half loop. It differs from the half-loop maneuver in that there is no transition into sustained straight and level inverted flight. It is a more complex maneuver than the half loop because there is no time to settle down inverted and pick up speed to roll upright. The accurate timing of the roll is critical to properly fly the Immelmann, and the aircraft is rolled at a slow airspeed, which requires exaggerated control movements and is less forgiving of errors.

Though this maneuver is named after Max Immelmann, the famous German World War I ace, his frail Fokker monoplane would have suffered structural failure attempting it. Immelmann most likely flew chandelles and wingovers to reverse direction. As the legend grew, the chandelles and wingovers slowly turned into a roll off the top of a half loop in the imagination. It seems that only the no-nonsense British are unmoved by the legend. They sensibly call this maneuver a *roll off the top*.

Typical minimum entry speeds
Aerobatic trainer: 160 mph
Intermediate aircraft: 160 mph
Unlimited monoplane: 180 mph

UNDERSTANDING IT

The forces acting on the aircraft in the Immelmann are exactly the same as in the half loop (*see* chapter 9) with one important exception. Because the roll off the top occurs the instant the aircraft reaches the top of the loop, the roll is done near the stall (especially in a low performance trainer), at a considerably slower airspeed than in the half-loop maneuver. Flawless execution of the slow roll is therefore crucial to avoid embarrassing sloppiness or even an inadvertent stall.

The same entry speed used for the half loop is sufficient for the Immelmann, but another 5–15 mph at the learning stage provides a bit of reserve energy if needed.

FLYING IT

The Immelmann is flown exactly in the same fashion as a loop or half loop up to the key point. At the risk of being somewhat repetitive, let's go all the way through the Immelmann.

1. Establish the proper entry speed in a shallow dive, and go into level flight momentarily, neutralizing controls. *Verify airspeed.* As in the loop, the stick will have to be slightly forward of neutral to maintain level flight because the aircraft is flying at a faster speed than for which it is trimmed.

2. Pull up into the loop. Be sure to gradually keep pulling the stick farther and farther aft because as the aircraft slows, the elevator becomes less effective and increasing back pressure is required to keep the loop's radius constant.

3. As the horizon disappears under the cowling, transfer your vision to the left wingtip, and, as in a full loop, watch for the wing chord to be 120° relative to the horizon (30° past the vertical). This is the key point.

4. At the key point, transfer your vision above your head to watch for the horizon and ease off the stick. As the nose approaches (descends toward) the proper straight-and-level inverted attitude, push the stick forward deliberately to stop the nose at the straight-and-level inverted attitude.

5. The instant the nose touches the appropriate inverted straight-and-level spot, briskly slow roll the aircraft upright. Review: For a left roll, apply full left aileron, simultaneously momentary right rudder followed by a good dose of left rudder, gradual aft stick toward neutral with about 30° to go, neutral aileron and rudder at the last moment, and the Immelmann is complete. Due to the aircraft's slow speed in the roll and the resulting deterioration in rudder effectiveness, you will have to apply more rudder than when rolling upright in the half-loop maneuver.

You should be able to roll to the right as well as to the left, so in the accompanying photo sequence a roll to the right is presented. See if you can figure out the appropriate control inputs and go fly it with your instructor.

Immelmann key list

- Line up over reference line
- Establish entry speed, level flight, *verify speed*
- Pull! Look left at wingtip as horizon disappears

- Keep pulling, vertical
- Key point, ease off stick, look ahead, up
- Nose on horizon, *roll* (upright)

COMMON ERRORS

Aside from the usual loop errors, applying excessive back pressure at the top of the loop (failing to sufficiently ease off aft stick) while rolling upright is the most common Immelmann error. In a high performance aircraft it can lead to a snap roll; if the aircraft snaps, there is little to do except immediately neutralize controls because when initiated, the snap roll completes itself and you will be right side up before you know it—perhaps impressing upon friends and sundry spectators (but not your favorite judge) that a snap roll was exactly what you wanted to do.

Some students find it difficult at first to remember to apply momentary rudder opposite the direction of the roll when commencing the roll upright. This oversight is perhaps because the roll is a smooth, integral part of the Immelmann and there is little time to think as the nose touches the horizon and the roll must be commenced. Failure to apply rudder results in barreling of the nose away from the longitudinal axis.

IF THINGS GO WRONG

By the time the student reaches the Immelmann stage, little should go wrong in the first half of the loop. An inadvertent snap roll at the top of the maneuver might be disconcerting but not unsafe, and it takes care of itself if controls are immediately neutralized. As is the case with any looping maneuver, the student should be also prepared to handle inadvertent inverted stalls, the dropping of the wing, and even an inadvertent spin for the rare occasions on which the controls are badly mishandled for whatever reason.

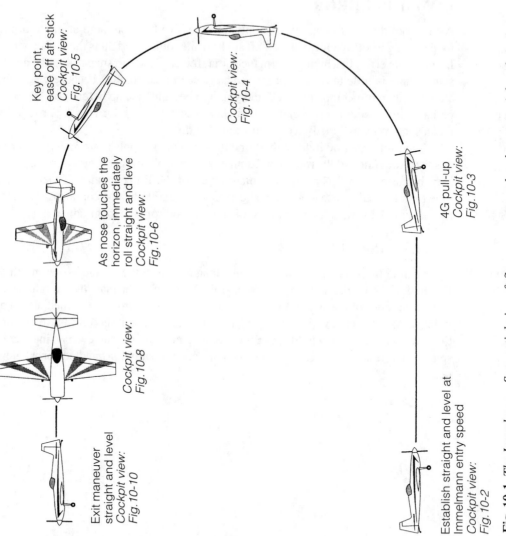

Key point,
ease off aft stick
Cockpit view:
Fig. 10-5

Cockpit view:
Fig. 10-4

4G pull-up
Cockpit view:
Fig. 10-3

As nose touches the
horizon, immediately
roll straight and leve
Cockpit view:
Fig. 10-6

Cockpit view:
Fig. 10-8

Establish straight and level at
Immelmann entry speed
Cockpit view:
Fig. 10-2

Exit maneuver
straight and level
Cockpit view:
Fig. 10-10

Fig. 10-1. *The Immelmann. Sequential aircraft figures correspond to the sequential cockpit views in the photos at the end of the chapter. The roll to straight and level is to the right.*

Fig. 10-2. *Establish straight and level at Immelmann entry speed.*

Fig. 10-3. *4G pull-up.*

Fig. 10-4. *Continue increasing aft stick as aircraft slows.*

Fig. 10-5. *Key point. Ease off aft stick, transfer vision forward to look for horizon.*

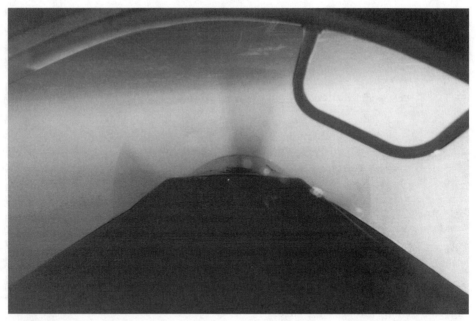

Fig. 10-6. *Roll instantly as nose of aircraft touches the horizon. (In this case, the roll is to the right.)*

Fig. 10-7. *Rolling toward knife-edge.*

Fig. 10-8. *Rolling through knife-edge.*

Fig. 10-9. *Rolling toward straight and level.*

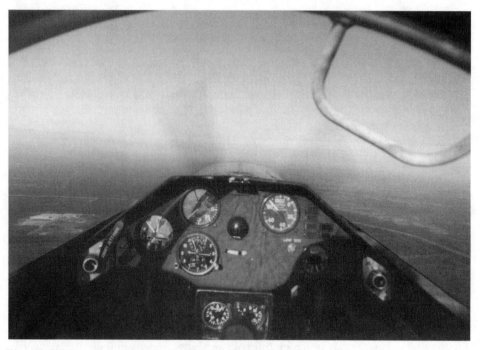

Fig. 10-10. *Straight and level exit from the maneuver.*

A great escape

One of the more dreaded scenarios in wringing out an aerobatic aircraft is suffering some form of structural failure. Staying within the aircraft's limits will probably always prevent such an event, but unanticipated component fatigue, faulty construction, or other unforeseen factors can occasionally cause in-flight structural failure. In such seemingly hopeless instances, quick thinking and a full understanding of all the forces acting on the aircraft might yet save the day. No incident illustrates this point better than Neil Williams' now legendary recovery from a lower wing spar attachment failure in a Zlin Akrobat.

Williams, a member of the British aerobatic team, was practicing for the 1970 world championships. He was not wearing a parachute. He was pulling five Gs at 1,000 feet when he heard a loud bang and felt a severe jolt. The aircraft started rolling to the left. The left wing was folding ominously upward. In spite of full right aileron and rudder, the roll to the left continued and the nose began to drop. At 300 feet, the aircraft was banked vertical and all control was about to be lost. For most pilots, it would have ended there.

But in much less time than it takes to write these lines, Williams recognized that the wing was folding under a positive G loading and a negative G load might just "unfold" it. He had no idea what the consequences would be, but it was the only option. He took it and applied full left aileron into the roll. As the Zlin rolled inverted into negative G flight, the wing snapped back into position with an almighty bang—audible to onlookers on the ground—and held. At this point the engine stopped. Williams had turned off the fuel, anticipating a crash. He quickly flipped on the fuel switch and the engine roared back to life.

From 150 feet agl, Williams gingerly began to climb and sort out what to do next. He had 8 minutes to decide, the amount of flying time for which the Zlin had sufficient fuel in inverted flight. His knees shook uncontrollably and, as he later said, he thought he was going to die. But he wasn't going without a fight and he already had a plan.

He would fly inverted as close to the airfield's grass surface as possible, and would roll upright at the last second. He even had the presence of mind to first experiment with rolling at altitude to see what effect the direction of roll had on the wing. He rolled left, saw the wing start to fold again and quickly returned to inverted flight. He made his inverted approach and rolled hard right at the last instant. The right wing cleared the ground by 6 inches. The left wing began to fold again but the aircraft settled into the ground upright with a resounding thump and as bits and pieces flew off in every direction, it slid to a stop. Williams was unhurt.

The accident investigation revealed what Williams realized as soon as the wing snapped back into position on the initial roll. The Zlin's wings are each held on by an upper and a lower wing spar attachment. The lower attachment failed as a result of repeated overstress during many hours of unlimited aerobatic flight. The upper spar attachment held and—together with Williams' immense experience, intricate knowledge of the forces acting on the aircraft, and a good dose of luck—it saved the day. Williams' subsequently stated that if he had been wearing a parachute, he would have climbed to altitude and jumped.

11
The Cuban eight

To UNDERSTAND THE CUBAN EIGHT, ONE MUST FIRST UNDERSTAND THE half Cuban eight. In the half Cuban eight, the aircraft completes ⅝ of a loop, is established on a line of flight at a 45° angle to the ground in the inverted position, is slow-rolled upright while continuing to maintain the 45° line of flight, and is pulled out into upright straight and level flight to complete the maneuver. The inverted and upright portions of the 45° line of flight must be equal in length, which means that the roll must be placed in the center of the 45° line.

The full Cuban eight is merely the immediate repetition of the half Cuban eight in the opposite direction. By doing the equal and opposite maneuvers the aircraft inscribes a horizontal figure eight in the sky.

Because the basic full Cuban eight is a repetition of the half Cuban eight, it is standard practice to teach only the half Cuban eight in basic aerobatics, which the student and instructor can then expand into the full maneuver at their own discretion.

The half Cuban eight (together with a variation, the reverse half Cuban eight) is among the most complex and challenging basic aerobatic maneuvers. It is a combination maneuver, utilizing elements of the loop, inverted flight, and the roll. During the Cuban eight, the aircraft ranges throughout the performance envelope, hovering near stall at the top of the loop, approaching redline on the pullout, experiencing negative G in the inverted phase, and building high positive G loads during the pull up into the loop and the pullout of the maneuver.

The half Cuban eight can be thought of as a more complex development of the half loop. In both maneuvers, the aircraft is transitioned into inverted straightline flight before being rolled upright. But while in the half loop, the transition is into inverted straight and level; in the half Cuban eight, the transition is into inverted 45° descending straightline flight with the added complications of a more challenging orientation requirement and the rapid buildup of airspeed.

The half Cuban eight is an important maneuver in aerobatic competition flying. The full Cuban eight—officially referred to as the *horizontal eight*—is rarely seen in competition in its basic form; however, complex variations of it are a staple element of advanced competition events. The full Cuban eight also remains one of the greatest airshow crowd pleasers of all time. It is especially graceful if done well in a big old biplane with a lot of smoke.

Typical minimum entry speeds
Aerobatic trainer: 140 mph
Intermediate aircraft: 160 mph
Unlimited monoplane: 170 mph

UNDERSTANDING IT

The basic aerodynamic forces experienced by the aircraft during the half Cuban eight have been extensively discussed in the sections covering its component maneuvers, the loop, inverted flight, and the roll; however, to fully understand the half Cuban eight, it is worth reviewing the forces at work during the whole maneuver, and noting some special implications.

The half Cuban eight can be started with the same amount of energy required for a loop because all we are trying to accomplish at first is to float the aircraft over the top as in any loop; therefore, airspeeds and acceleration forces are the same as in the ordinary loop.

Airspeed builds rapidly and moderate negative G is experienced as the aircraft begins to descend in the second half of the loop and is established in straightline inverted flight along a 45° descending line.

The roll upright must be accomplished promptly, at maximum roll rate, as speed continues to build. The technique is similar to that of the slow roll, with one exception: the elevator is kept neutral. As the airspeed increases, the wing generates progressively more lift that compensates for the loss of lift experienced in a straight and level slow roll, and eliminates the need for forward stick.

Considerable acceleration load is generated during the pullout from the half Cuban eight, when the upright segment is completed.

The critical segment of the half Cuban eight is the 45° descending straightline segment. Much needs to be accomplished as airspeed builds rapidly, and initially the view from the cockpit can be quite disorienting. Errors can quickly send the aircraft into the performance envelope's upper limits. Too much time spent rolling upright, or a descent angle steeper than 45°, will result in excessive speed.

A panicky pullout at high speed can send the G meter soaring. While learning the maneuver it is not uncommon to reach redline on the pullout. So it is important to understand the aerodynamic forces acting on the aircraft at all stages of the half Cuban eight; to be timely and precise in performing the maneuver; and to understand and be willing and able to promptly apply all recovery options if at any point the maneuver exceeds parameters.

FLYING IT

The half Cuban eight presents a higher workload than most basic maneuvers. It is a complex maneuver, so it is especially important to be mentally well ahead of the airplane, to anticipate each segment. Don't be discouraged if it takes awhile to get the hang of the half Cuban eight. When you master its subtleties, you will find it among the most rewarding aerobatic maneuvers in your repertoire.

Following the detailed description of flying the half Cuban eight, an abbreviated key steps list is also provided. Many pilots find it helpful to mentally check off these steps as they fly the half Cuban eight. Let's see how it is flown.

1. Line up over a road for the loop and increase airspeed to the aircraft's loop entry speed. Establish momentary straight and level flight, and neutralize the controls. *Verify airspeed.* You will need slightly forward stick to maintain level flight at the higher speed because you didn't (and shouldn't) retrim for the entry speed.

2. Begin a steady 3.5G–4.0G pull-up by applying smooth back pressure on the stick. Glance briefly at the G meter to confirm the acceleration rate. Be sure to apply steadily increasing back pressure to the stick on the way up to keep the radius of the loop constant as the elevator becomes less effective with the decreasing airspeed.

3. As soon as the horizon disappears under the cowling, be sure to look out at the left wingtip to monitor the loop's progress.

4. At the key point—120° into the loop (30° beyond vertical)—look straight up to watch for the horizon to appear, ease off slightly on the stick and float gently over the top of the loop.

5. As the aircraft goes over the top, pick up the horizon and then the reference line below on the ground to check how accurately the aircraft is aligned in the maneuver.

6. You now have to quickly ascertain the instant when the aircraft reaches the point where its longitudinal axis is at 45° to the ground, and establish straight-line flight along this line. There are two ways to establish the 45° line. The precise method, to be used regularly as you become comfortable in the maneuver, is to look out at the left wingtip and wait for the wing chord line to form a 45° angle with the horizon.

   Some people find this method too distracting at first, and for them there is a less precise method to be used in the initial stages of learning the maneuver. Continue to look forward as you float over the top of the loop. As the nose reaches the horizon, pick a prominent spot on the ground approximately half way between the horizon and the point straight under the aircraft. When the nose reaches that spot, for training purposes you can use it as the 45° mark.

7. When the aircraft reaches the 45° mark, briskly push the stick forward to hold the nose on it and establish straightline flight along this incline. It is important that the aircraft fly a short but perceptible distance in straightline inverted flight, tempting as it might be to start rolling upright immediately. To avoid the temptation, count "one thousand, two thousand." You will then be at the proper spot to roll upright.

8. Briskly execute a slow roll upright. It is important to roll quickly with full aileron deflection because your speed is increasing and if you take your time rolling it might become excessive. In the roll, you will need the usual rudder in the direction of the roll as you go past vertical (top rudder), but neutral elevator will be sufficient instead of forward stick. As airspeed increases, so does lift generated by the wing that will be sufficient to keep the nose on the point around which you are rolling.

9. Once upright, you must continue in descending straightline flight along the 45° incline for the same distance that you flew in straightline inverted. So, count again "one thousand, two thousand"—say "two thousand" a little faster because of the increasing speed—and pull out into straight and level to complete the half Cuban eight. Because you will be at a high speed and might well be approaching redline, be gradual on the pullout, keep an eye on the G meter, and be sure not to exceed limit load.

   Always keep your hand on the throttle and if necessary throttle back *immediately* anywhere along the 45° straightline segment. Make it a "no brainer" yes-no decision, established in advance; select an appropriate maximum speed for this segment, and the moment the ASI needle reaches that maximum speed, throttle back—as simple as that with no fudge factor.

Cuban eight key list

- Line up on reference line
- Loop entry speed, level flight, *verify speed*
- Pull! 4Gs, eyes left
- Key point! Ease stick forward, look up
- Horizon, check alignment with ground reference line; look left
- 45°! Forward stick, look forward, stop nose
- One thousand, two thousand, roll hard!
- One thousand, two thousand, pull out

COMMON ERRORS

A number of common errors that can occur during the half Cuban eight will result in a poorly flown maneuver.

Failure to establish a 45° down line, failure to neutralize aft stick prior to rolling. The ground looms large in the windscreen and gets even larger at an alarming rate even at a very respectable altitude. The temptation for the novice to roll as soon as possible is great, so aft stick is often not neutralized and the 45° downline is not established. The consequence is that the aircraft is not moving in a straight line when the roll is begun and comes off heading in the half roll to upright, doing a sort of half barrel roll. Coming off heading by as much as 90° is not unheard of.

Application of aft stick during the slow roll to upright flight, even though the 45° downline has been established. In the hectic environment of the final stage of the half Cuban eight, it is easy to get distracted and apply some aft stick in the roll to straight and level. The results are, again, a sort of half barrel roll and completion of the roll off heading.

Failure to keep rolling. As the airspeed increases on the 45° downline, the stick forces also increase. What felt like a normal amount of aileron control input in a straight-and-level roll is equal to practically no aileron at this stage of the half Cuban eight, and the aircraft might even stop rolling. Not a time to dillydally when the error is recognized, but to roll briskly, and perhaps even retard the throttle if the airspeed rushes uncomfortably toward redline.

IF THINGS GO WRONG

Alarmingly increasing airspeed, a rapid loss of altitude, and the threat of overstressing the aircraft in a high-G pullout are the bugbears of the half Cuban eight, all threatening on the 45° downline segment of the maneuver. Three points should be embedded in your mind for handling unraveling half Cuban eights on this segment:

- Reduce power!
- Pull up to avoid exceeding V_{ne} but at a rate that keeps you well within G limits.
- Always perform the half Cuban eight at an altitude from which running out of altitude is not even a potential problem.

If you inadvertently stopped rolling on the 45° downline, reestablish the roll to recover to upright flight. *Do not under any circumstances panic and try to pull through in a split S to upright flight!* The excess speed you would build in a split S, the altitude you would lose, and the Gs you might pull to recover would far exceed anything you will encounter rolling upright and could have catastrophic consequences.

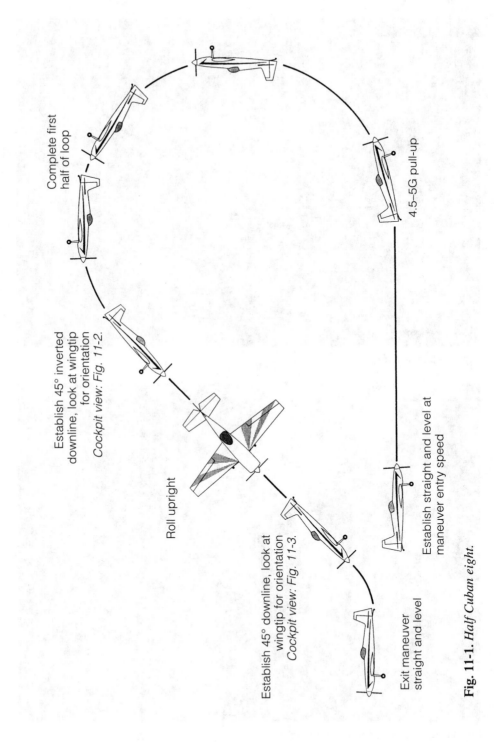

Fig. 11-1. *Half Cuban eight.*

Fig. 11-2. *Inverted 45° downline.*

Fig. 11-3. *Upright 45° downline.*

Patty Wagstaff flying her Extra 300S in which she won her third consecutive unlimited U.S. national championship in 1993.

The Sukhoi SU-26MX (top) and SU-29 are a formidable state-of-the-art unlimited aerobatic pair.

The two-seat Christen Eagle, among the best of the kit-built machines, and a joy to fly.

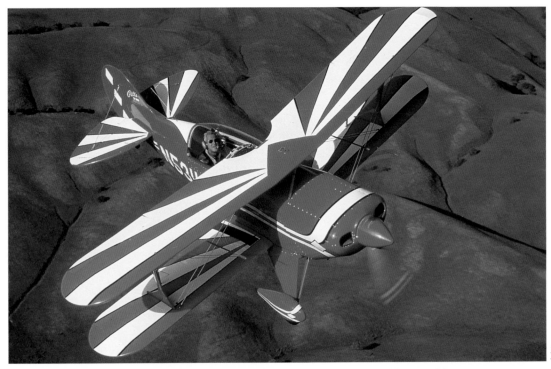

The Pitts Special S1-T, an eternal classic and still one of the great competition machines.

Coauthor Mike Goulian in his Staudacher 300GS that won him a spot on the U.S. National Aerobatic Team in 1993.

With an inverted system, you can enjoy this new perspective offered by aerobatic flight, at your leisure.

The Canadian Snowbirds make a mass aerobatic formation act look so easy.

Smoke's on! Mike Goulian perfecting his airshow routine in the Staudaucher 300.

12
The reverse Cuban eight

THE REVERSE CUBAN EIGHT IS REALLY THE REVERSE HALF CUBAN EIGHT flown twice in succession in opposing directions. It is therefore standard practice to teach only the reverse half Cuban eight, which can then be expanded at the student's discretion into a full reverse Cuban eight.

In the reverse half Cuban eight, the aircraft is pitched up 45° from straight and level flight and established in straightline flight along this incline. It is then rolled inverted and kept in straightline flight along the 45° incline for the same distance it flew upright along this segment. It is now in a position equivalent to the key point of a loop. From this point, the aircraft is floated over the top into a descending half loop to complete the reverse half Cuban eight.

The reverse half Cuban eight combines elements of the roll, inverted flight, and the loop. It is a fundamental competition maneuver. While the full reverse Cuban eight is seldom flown in competition, it remains a popular element in airshow display flying.

Typical minimum entry speeds
Aerobatic trainer: 160 mph
Intermediate aircraft: 160 mph
Unlimited monoplane: 180 mph

UNDERSTANDING IT

The aerodynamic forces acting on the airplane during the reverse half Cuban eight have been covered in the sections on its component elements, the roll, inverted flight, and the loop. What makes the reverse half Cuban eight an interesting challenge is that during the maneuver, the aircraft is flown through a wide range of the performance envelope and at times might approach its limits.

Entry speed for the reverse half Cuban eight must be high because a lot of energy is required on the ascending 45° straightline segment. In fact, among basic aerobatic maneuvers, the reverse half Cuban eight has the highest entry speed. About as much positive G is pulled on the aggressive pull up to 45° as is pulled upon entering a loop.

The airspeed diminishes fairly rapidly on the ascent, so the roll has to be prompt and at the maximum roll rate. The roll to inverted is vintage slow roll, forward stick being necessary to keep the nose up due to diminishing airspeed, which results in diminishing lift. The 45° upline has to be precisely that. If it is more, the aircraft will slow prematurely and could stall while rolling; if it is less, the aircraft will not slow down at the top of the maneuver and will gain excess airspeed on the half loop down.

In the inverted position, the aircraft is subject to mild negative G. As the transition into the top of the looping segment is accomplished, the aircraft will be at the slowest airspeed and will approach the critical angle of attack. The airspeed increases rapidly, and high positive G builds as the aircraft is pulled through to complete the half loop. In the learning stage, the ASI might approach redline on the pullout.

The most critical part of the reverse half Cuban eight is the completion of the half loop. If it is misjudged and started prematurely from too fast an airspeed, it can result in a tremendous loss of altitude, speed beyond redline, and excess G loads on a panicky pullout. The downward half loop is sometimes referred to as the *split-S*.

Because of the potentially catastrophic hazards of even a small misjudgment, especially at low altitudes, the split-S is not presented as a separate maneuver. It is the authors' strong opinion that the so-called split-S *should not be flown under any circumstances in basic recreational aerobatics except during dual training* to demonstrate its folly, and as a segment of the reverse Cuban eight when it is entered into at an extreme nose-high attitude, making the required very slow entry speed easily manageable.

FLYING IT

The reverse half Cuban eight is a lyrical maneuver when flown well. As the control movements become second nature, the rush of speed, the effortless roll around the horizon, the practically motionless pause at the top, and the sweeping, accelerating arc back to straight and level becomes an exhilarating aerial ballet, aerobatics at its best. Let's see how it is flown. A key steps list follows the detailed description.

1. Line up over a road or other straightline landmark as if preparing for a loop. Dive slightly to establish the proper reverse half Cuban eight entry speed. Establish momentary level flight, *verify airspeed*.

2. Aggressively pull up at a 3.5G–4.0G acceleration rate. Look at the left wingtip and wait for the wing chord to reach 45° relative to the horizon.

3. Briskly establish straightline flight along the 45° incline with neutral stick;

look ahead and count "one thousand, two thousand" to stay on the established flight path for an identifiable distance.

4. Rapidly roll into the inverted position. Use the standard slow roll technique, and be sure to apply some forward stick (but not too much; refer to the horizon) to counter the lift being lost as the aircraft slows and the nose wants to sink toward the horizon. Remember, you must continue along the 45° incline, so the nose will be way above the horizon compared to what you are used to seeing in straightline inverted flight.

5. Maintain straightline inverted flight along the 45° degree incline and count "one thousand, two thousand," glance at the ASI and if your timing was correct throughout the maneuver, you should now be hovering near the published stall speed, about to float over the top of a loop. If you are still too fast, let the aircraft continue along the straight line and dissipate airspeed to just above the published stall speed.

6. Now all that remains is to transition into the loop, float over the top, and continue down the back to upright straight and level flight. Ease the stick back slightly toward neutral from its forward position and float gently over the top. It cannot be emphasized enough that you must *ease* off *slightly*. Do not pull on the stick; let the stick pressure bring it back. If you pull on the stick, you will initiate the back end of the loop prematurely and will slice off the perfect circular descent path you are aiming for.

   It also cannot be emphasized enough that you must initiate the looping segment only at the proper slow speed. Any excess speed at the top will result in an alarming airspeed buildup on the way down. As little as an excess 10 mph at the top can send the ASI to redline and even beyond in inexperienced hands.

7. As the aircraft begins its downward trajectory and begins to build speed, apply increasing back pressure on the stick as you would in any loop, and level out at the bottom to complete the maneuver.

Reverse half Cuban eight key steps

- Entry speed, straight and level, *verify entry speed*
- Pull! 4Gs, eyes left
- 45°, stick neutral, eyes forward, straightline flight
- One thousand, two thousand, roll hard!
- Inverted straightline, one thousand, two thousand, keep nose up
- Check airspeed, *ease* off stick toward neutral
- Float over the top; look at the ground reference line, check alignment
- Let the aircraft start down, then pull, gradually. Check ASI and G meter on the way down. Keep pulling

COMMON ERRORS

The common errors encountered during the reverse half Cuban eight not only affect the quality of the maneuver, but can also pose a threat to safety.

Establishing an upline less than 45°. If the upline is less than 45°, the aircraft will fail to slow down sufficiently to properly complete the half loop down. Were the maneuver to be continued, an enormous increase in airspeed and loss of altitude would result with the possibility of overstressing the aircraft on pull out. The only reasonable course of action is to abandon the maneuver, roll upright, and start over.

It is difficult for the novice to accurately establish the 45° upline because at first it looks so much steeper than it is. Practice and a good instructor are the cure.

Establishing an upline greater than 45°. Some enthusiastic pilots err on the other side. In this case the aircraft slows too fast and might stall during the roll.

Failure to sufficiently slow down and allow the aircraft to float over the top. Many students are anxious to get on with the maneuver when the 45° upline segment is completed, and seem not to have the patience to allow the aircraft to properly run out of kinetic (airspeed) energy and gently float over the top. The airspeed increase and altitude loss during the half loop down would be excessive, so it is best to abandon the maneuver, push back into inverted straight and level flight, roll upright, and start over.

IF THINGS GO WRONG

The three things that can most frequently compromise safety during the reverse half Cuban eight are the excessive buildup of speed, excessive altitude loss, and excess airframe stress on pullout, all on the half loop down segment. The causes, as pointed out above, are an upline less than 45° or a failure to allow the aircraft to properly slow down and float over the top of the maneuver. In both instances, the cure is to *abandon the maneuver early and roll upright* instead of pulling through to upright straight and level.

Excessive airspeed buildup and altitude loss might also occur in less extreme fashion if the 45° upline segment is properly flown and the aircraft is floated over the top, but the half loop down is mishandled by failing to pull hard enough soon enough. In this case:

- *Reduce power!*
- Pull up to avoid exceeding V_{ne}, but at a rate that keeps you well within G limits.

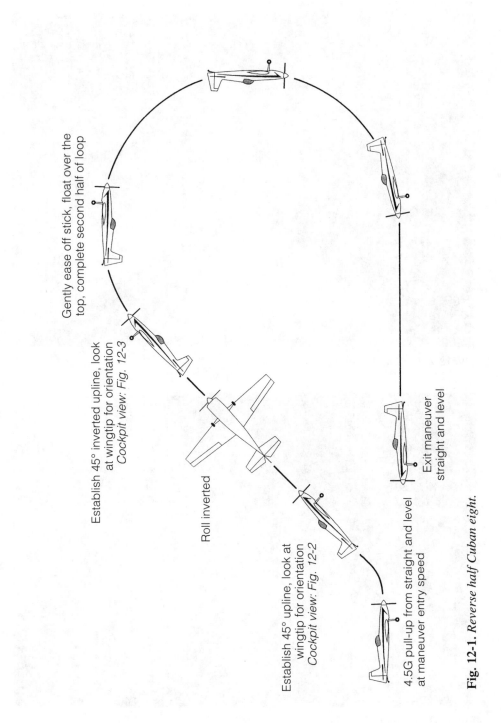

Gently ease off stick, float over the
top, complete second half of loop

Establish 45° inverted upline, look
at wingtip for orientation
Cockpit view: Fig. 12-3

Roll inverted

Establish 45° upline, look at
wingtip for orientation
Cockpit view: Fig. 12-2

4.5G pull-up from straight and level
at maneuver entry speed

Exit maneuver
straight and level

Fig. 12-1. *Reverse half Cuban eight.*

Fig. 12-2. *Upright 45° upline.*

Fig. 12-3. *Inverted 45° upline.*

Mustang dogfight

"The Focke Wulf FW 190 is level with me at 11 o'clock, 3 miles, closing fast, almost head on. The guns are armed, I advance the Mustang's power to the military setting and tense up for the encounter. The 190 will outturn me, but I can outrun him any time. Force him vertical. We'll need all the speed we can get. I unload the Mustang with forward stick for best acceleration at zero G. The 190 matches the shallow dive, I skid to put more lateral distance between us to turn into angular distance in the vertical plane for more room to maneuver . . . *he flashes past*, hard back on the stick, 5Gs, look back at him, *never lose sight*! Yeah, he breaks left, pull down on him, rudder the nose after him! He's reversing, turning right! Roll to the left! We're closing, he's in the gun sight . . . lead him . . . a little more . . . *fire*!"

True story, but not World War II, and the Focke Wulf was a ghost. The flight took place in the Stallion Corporation's rare full dual-control TF-51 Mustang, with Doug Schultz, co-owner and former Navy F-4 Phantom pilot, who with his partner offers a series of rental programs in the Mustang from novice introductory flights to a full checkout.

A couple of interesting points: at slow airspeeds aerobatic maneuvers in the Mustang are a dream, but faster than 300 knots, the controls stiffen considerably. Our "dogfight" took all of about 1 minute and we were down to 1,000 feet from 10,000 feet. When we were pointed down at the imaginary Focke Wulf, we were not going nearly as fast as I had expected. Our maneuvering used up so much energy that we were only at 250 knots, about where the Focke Wulf would have been after its tight reversal. And, yes, the flight was a dream come true!

13
The hammerhead

DURING THE HAMMERHEAD THE AIRCRAFT IS PULLED UP INTO A VERTICAL climb and established in straightline flight along the vertical upline. It is allowed to fly along the up line until the instant before it runs out of airspeed (kinetic energy) and almost becomes stationary. It is then pivoted 180° around its vertical axis and established in straightline flight along the vertical down line parallel and opposite to the upline. It is allowed to gain speed along the downline and is then pulled out into straight and level flight opposite to the entry direction, to exit the maneuver.

During the hammerhead, the aircraft approaches the extremes of the performance envelope in terms of limit load and airspeed. It is an important competition maneuver. In competition, the length of the upline and downline do not have to be equal (respective length will depend on the preceding and following maneuvers), though they both have to be clearly established.

Typical minimum entry speeds
Aerobatic trainer: 160 mph
Intermediate aircraft: 180 mph
Unlimited monoplane: 190 mph

UNDERSTANDING IT

The hammerhead is an interesting maneuver because three forces, slipstream, torque, and gyroscopic precession caused by the aircraft's propeller come into play to a greater extent than in most other aerobatic maneuvers. The effect of these forces is especially pronounced in high-performance competition aircraft.

The beginning of the hammerhead is similar to the beginning of a loop; it can be thought of as performing a quarter loop, which would place the aircraft on the vertical upline. The minimum speed for a hammerhead is the loop entry speed, though the faster the entry speed, the longer the vertical upline, and the more well-defined and distinctive the hammerhead. In a Decathlon, a good entry speed is 160 mph, or about the same airspeed used for the reverse half Cuban eight.

The interesting forces become apparent as the aircraft begins to run out of energy along the upline. The slipstream that is usually drawn out behind the aircraft becomes tighter and tighter as the aircraft slows. The slipstream forces the fuselage and tailplane to the right as it tightens, which causes the left wing to drop, so it has to be countered with slight right rudder.

As the aircraft slows on the upline, the wing produces progressively less lift. There comes a point when the engine torque effect overcomes the wing's lifting effect and causes the aircraft to roll left. This has to be countered with slight right aileron.

As the aircraft becomes practically motionless, the slipstream becomes so tight around the fuselage that it buffets the airframe, much as it does on the ground during engine runup. This usually is the signal to begin to pivot with the rudder.

As the nose rotates when the aircraft pivots, there are two forces to contend with. The outside wing generates more lift than the inside wing because it is traveling faster in the pivot and gyroscopic precession is created by the twisting from right to left of the rotating propeller disk. The greater lift on the outside (right) wing wants to roll the airplane left, and gyroscopic precession wants to push it over on its back. Aileron opposite to the direction of the pivot overcomes the outside wing's extra lift, and slight forward stick counters gyroscopic precession.

A word of caution on this control configuration, especially for pilots of all Pitts and Eagle aircraft. Forward stick and full rudder deflection coupled with high power and negligible airspeed is the recipe for an inverted flat spin, and if forward stick is excessive, the aircraft will spin in a flash. In aircraft susceptible to this phenomenon, it is imperative to obtain expert instruction in the recognition of and recovery from the inadvertent inverted spin from a hammerhead, and to manage altitude conservatively.

Care must be taken not to exceed redline and G limit on the pullout after the aircraft has been established in straightline flight on the vertical downline and has built up speed.

FLYING IT

The hammerhead is not a particularly difficult maneuver to fly. Rather than demanding finesse it only requires the mechanical application of the appropriate control inputs at the appropriate moments to look right. Let's see how it is flown. In our example, the aircraft will pivot to the left at the top of the hammerhead.

1. The first order of business is to find a straightline ground reference, such as a road or power line. You will fly the hammerhead perpendicular to the road, so

that you can look at it for reference on the way up, during the pivot, and the vertical down segment.

2. Establish hammerhead entry speed as you set up a flight path perpendicular to the road. At the entry speed establish momentary straight and level flight and neutralize the controls (the stick will have to remain forward of neutral because of the greater than trimmed for airspeed). *Verify airspeed.* Time the straight and level segment to take you directly over the ground reference line at the correct speed.

3. Aggressively pull up to establish the aircraft on the vertical upline flight path. You have to pull slightly more Gs than when flying a loop to achieve a crisp, solid transition to vertical flight.

4. As the horizon disappears under the nose, look at the left wingtip to monitor your position in the maneuver. When the wing chord line is perpendicular to the horizon, apply forward stick toward neutral to establish vertical straight-line flight.

   Your reference is the wingtip's position on the horizon, and your objective is to keep it stationary in its proper position relative to the horizon. Two controls are used to maintain position:
   ~ Rudder moves the wingtip up and down relative to the horizon (controlling yaw).
   ~ Aileron moves the wingtip backward and forward along the horizon (controlling roll).

5. As the aircraft slows on the way up and nears the top of the hammerhead, the slipstream tightens around the fuselage and might force the fuselage and tailplane right, requiring slight right rudder pressure to counter it. Your visual clue of this effect is a movement of the left wingtip downward relative to the horizon.

   The wings produce less and less lift as the aircraft slows and at some point the torque force might exceed the effect of lift rotating the aircraft to the left along its longitudinal axis. Your visual clue of this effect is a movement of the left wingtip forward relative to its initial position on the horizon. Right aileron counters this effect and maintains the wingtip in its proper position.

   Slipstream and torque effects are most noticeable in high-performance aircraft with small wings, such as the Pitts. In a training aircraft such as a Decathlon, the effect is too small to worry about during the initial stages of learning the maneuver.

6. The next phase of the hammerhead is the pivot. It requires the systematic sequential application of all three controls. As the aircraft almost comes to a standstill, the tightening slipstream mildly buffets the fuselage. The instant this happens, brisk full rudder must be applied in the desired direction of the pivot. This example pivots left; thus, you apply full left rudder.

As the aircraft pivots, the outside wing (in this example, the right wing) will travel faster than the inside (left) wing, generating more lift and rolling the aircraft left as it pivots. It is countered by full opposite (right) aileron.

In an aircraft with a propeller that rotates to the right, gyroscopic precession forces the nose of the aircraft toward the pilot during the pivot, tending to push the aircraft over on its back. It is countered with slight forward stick.

Thus the pivot requires three distinct sequential control inputs. It is imperative to perform them in sequence rather than simultaneously. You will find the procedure a mechanical exercise, requiring little finesse. To review, for a left pivot:

- ~ First, brisk full left rudder
- ~ Second, full right aileron
- ~ Third, slight forward stick

A note about "brisk rudder." It does not mean kicking the rudder. It is a swift, deliberate, but controlled push, rather than the lightning-flash fury of a kick. Very few things in life need to be kicked and an aircraft rudder is never one of them.

7. Now you look straight ahead and monitor the progress of the pivot. As the nose passes through the horizon, glance at the ground. The spot at which the wing (now vertical relative to the ground) points is the spot where the nose should point when it reaches the vertical downline.

8. As the nose reaches the point 45° short of the vertical downline, briskly apply full momentary opposite right rudder and simultaneously, in one continuous process, neutralize all controls. When the controls are neutralized, the aircraft should be on the vertical downline heading straight for the spot you saw along the wing when the nose passed through the horizon during the pivot.

9. Allow the aircraft to fly along the vertical downline to establish a recognizable straightline flight path along the vertical downline, and then pull out into straight and level flight as you would from a loop. To achieve a crisp transition, pull about as many Gs as you did going into the hammerhead. When you exit the maneuver, you should be going 180° opposite to your entry heading.

Hammerhead key list

- Establish entry speed; over the road; *pull!*
- Look left; vertical, neutral stick
- (Aircraft buffets) Left rudder, right aileron, slight forward stick in sequence
- (Nose goes through horizon) Glance left, note spot, look ahead
- 45°, full right rudder, neutral controls
- Straight down . . . wait . . . *pull!*

COMMON ERRORS

Errors during the hammerhead are most likely to occur during the pivot, which is the most unfamiliar segment of the maneuver for the aerobatic student, but they might also be committed during other stages. Let's look at them in sequence.

On the pull-up to vertical, a line shallower than vertical is established. You are not paying attention to the wing chord line's angle with the horizon, and are establishing straightline flight too early.

The time to pivot isn't recognized in time. The slipstream's buffeting of the fuselage is subtle and gentle. Watch for it carefully.

The pivot is put off so long that the aircraft starts to tailslide back down in the direction from where it came. If it starts to tailslide, it is too late to continue with the hammerhead because the reversed direction of the airflow over the airframe might cause the airplane to behave differently than required in the hammerhead. At this stage, the maneuver should be aborted. This is accomplished by pulling the stick full aft and firmly holding it there. Alien as this might sound, it is the standard technique for performing a forward tailslide, which is the advanced aerobatic maneuver you are now inadvertently flying. All that will happen is that the aircraft will arch forward into a dive. When the nose points straight down, release back pressure and recover from the dive.

Rudder, aileron, and forward stick are applied simultaneously at the top of the pivot. This error will cause the aircraft to go off proper alignment in the maneuver. If you have trouble with sequential control application, verbally call out each control in the proper order and apply each control as you say its name.

The controls aren't properly neutralized at the 45° point prior to the vertical downline. This error will cause the aircraft to be misaligned and off heading on the pullout. In some aircraft, at this point of the maneuver, forward stick and some rudder in either direction can cause an inadvertent inverted spin, which is covered in the next subheading of this chapter. Take great care to neutralize the controls.

The aircraft is pulled out of the hammerhead after the pivot without being first flown in straightline flight along the vertical downline. This usually happens because the pilot is concerned with building up too much speed and is rushing the maneuver. Bear in mind that when you start heading down you are at zero airspeed. You have a reasonable amount of time to establish a vertical downline before you pull out.

Excessive speed builds up on the downline. Just as you should not pull out too soon, you also shouldn't let too much speed build up on the way down. Be ready to retard the throttle, and be careful not to overstress the airplane on pull out, which is covered in the next subheading.

IF THINGS GO WRONG

Generally, four scenarios might threaten the safety of the flight if not properly recognized and handled:

- An inadvertent tailslide at the top of the hammerhead
- An inadvertent inverted spin because of a failure to neutralize controls while transitioning to the vertical downline
- Excessive speed buildup on the vertical downline
- Pulling excessive Gs on recovery

As described among common errors above, the inadvertent tailslide might occur if the pilot fails to initiate the pivot in time at the top of the hammerhead. It might threaten safety if not properly handled because the reversed airflow over the fuselage could cause the aircraft to enter a maneuver unfamiliar to the pilot with basic aerobatic skills, cause disorientation, and make it difficult to recover.

As described in common errors, the cure for the tailslide, which is an advanced aerobatic maneuver, is to simply complete it according to standard procedure. Pull the stick full aft, keep the rudders neutral, and wait. The aircraft will go over on its back into a dive from which recovery is simple.

Completion of the inadvertent tailslide is really a benign maneuver and should be practiced while learning the hammerhead.

An inadvertent inverted flat spin situation can develop in certain aircraft, especially the Pitts, Eagle, and some other high-performance aircraft, if rudder and elevator are not properly neutralized during the transition to the vertical downline, or if excessive forward stick is applied during the pivot. The control input for the inverted spin is forward stick and rudder in the desired direction of rotation.

So, if you are pointing straight down and carry forward stick and right rudder because you neglected to neutralize them, you might find yourself in an inverted flat spin. It is not at all dangerous if you know how to recognize it and what to do. (For specific recovery techniques, see your flight manual.) Power off, rudder opposite to the direction of yaw, and neutral stick will put you back into a straightline dive. (See chapter 14.)

If your aircraft is susceptible to inadvertent inverted spinning under the described conditions, include the appropriate recovery training in your hammerhead curriculum.

Excessive speed buildup on the downline should be a rare problem. You transition to the downline at practically zero airspeed; however, if speed does build excessively, retard the throttle immediately, and simultaneously start your pullout. Eye the G meter to stay within limits.

Excessive G buildup on pullout might be related to having built up excessive speed, or you might just be too enthusiastic about pulling out. There is really no cure, only prevention. Focus consciously on G load throughout the pullout and you will stay within limits. If you notice the problem too late, ease off the stick immediately.

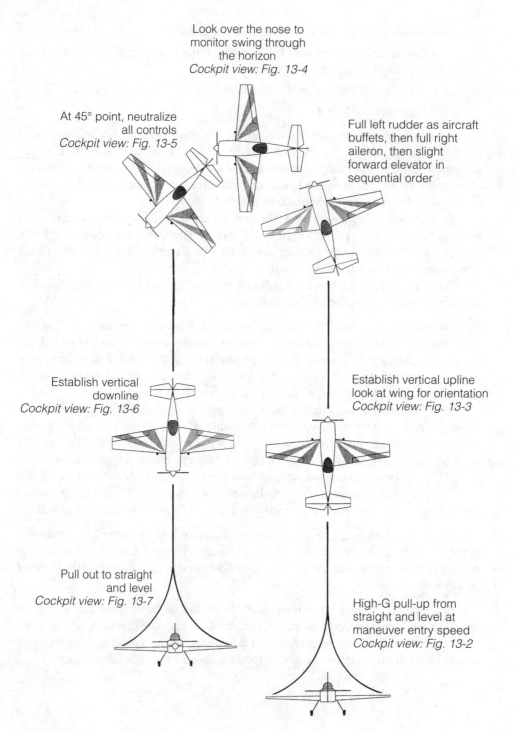

Look over the nose to
monitor swing through
the horizon
Cockpit view: Fig. 13-4

At 45° point, neutralize
all controls
Cockpit view: Fig. 13-5

Full left rudder as aircraft
buffets, then full right
aileron, then slight
forward elevator in
sequential order

Establish vertical
downline
Cockpit view: Fig. 13-6

Establish vertical upline
look at wing for orientation
Cockpit view: Fig. 13-3

Pull out to straight
and level
Cockpit view: Fig. 13-7

High-G pull-up from
straight and level at
maneuver entry speed
Cockpit view: Fig. 13-2

Fig. 13-1. *The hammerhead.*

Fig. 13-2. *High-G pull-up.*

Fig. 13-3. *Establish vertical upline. Look at wing for orientation.*

Fig. 13-4. *Look over the nose to monitor alignment through the pivot.*

Fig. 13-5. *At the 45° point, neutralize all controls.*

Fig. 13-6. *Establish vertical downline.*

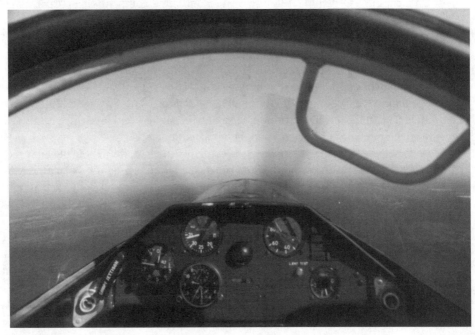

Fig. 13-7. *Pull out to straight and level.*

Fixed "constant-speed" propeller

If your airplane has a constant-speed propeller, setting the power for aerobatic flight is easy. You set the prop control and throttle to a high power setting (2,500 RPM and 25 inches in a Decathlon, full power in an unlimited competition aircraft), and off you go, confident that the propeller will not overspeed. But a lot of aerobatics is flown in aircraft that are not equipped with a constant-speed propeller. What to do, then, to prevent prop overspeed without having to constantly pump away at the throttle?

A very effective technique allows the aerobatic pilot flying with a fixed-pitch prop to set power once, and leave it alone for most of the aerobatic session. It is known as "redlining the tachometer on the airspeed indicator." The way to do it is to go up to altitude, open the throttle to full for aerobatic maneuvers and dive the aircraft, watching the tachometer (RPM gauge) and the airspeed indicator.

When the tachometer reaches its acceptable limit (redline), note the airspeed that is registered on the airspeed indicator. This is the redline speed for the engine RPM. You can safely fly to this speed at the present throttle setting without exceeding the engine's RPM limit. Mark this speed prominently on the airspeed indicator and you are in business.

All you have to watch during aerobatic sessions is the airspeed indicator. If tach redline speed is about to be exceeded, throttle back slightly, and when the aircraft slows, reset the throttle to its original full-throttle position. You will find that this technique yields minimal need for power adjustments.

14
Spins

AT LAST WE COME TO THE SPIN, THE MANEUVER MOST FEARED BY PILOTS with little or no spin experience. Why learn it so late? Because by learning all the other maneuvers first, many of which are far more extreme in unusual attitude, acceleration, and airspeed variation, you will find that the dreaded spin is not nearly as fearsome as it is made out to be by the uninformed. You will also realize by the time you come to learning the spin that falling into an inadvertent spin during most basic aerobatic maneuvers in most aircraft (with the possible exception of the hammerhead) is not a frequent occurrence.

While you will have learned many basic aerobatic maneuvers by the time you get to the spin, it is imperative that before doing any solo aerobatics for the first time, you receive thorough dual spin instruction from a qualified instructor.

The spin is an autorotational maneuver in which the aircraft is initially yawed and stalled, and as a result drops a wing. The inside wing is fully stalled while the outside wing continues to generate some lift and rotates around the inner wing as the aircraft descends in a stabilized condition.

The Flying It subsection of this chapter primarily addresses upright competition spins, with another subsection that addresses emergency spin recovery techniques. The final subsection briefly addresses recovering from inadvertent inverted spins. (A deliberate inverted spin is an advanced maneuver beyond the scope of this book, but inverted spin recovery is best learned by all pilots performing aerobatics at any level.)

UNDERSTANDING IT

To spin, an aircraft must first stall. A spin results when in a stalled condition a wing drops as the aircraft is deliberately or inadvertently yawed with rudder or by the gyroscopic force of the propeller (in a hammerhead and similar low-speed maneuvers, the main cause of inadvertent spins is the propeller's gyroscopic force). As a result of the stalled and yawed condition and the associated relative wind, the inner (slower) wing has a higher angle of attack than the outer (faster) wing. The inner wing fully stalls and drops into a bank while the outer wing continues to generate some lift. The outer wing's lift keeps it flying, rotating around the fully stalled inner wing as the aircraft descends. This is autorotation—a spin.

Spins may be entered deliberately or inadvertently. In an inadvertent spin entry, the aircraft—while being maneuvered for another purpose—is unintentionally flown in an uncoordinated fashion, which triggers a spin.

No spin is inherently dangerous, but the loss of altitude in a spin is greater than in any other basic aerobatic maneuver. The greatest danger in the spin is running out of altitude before being able to recover. Typical inadvertent spin accidents occur when at low altitude and slow airspeed, such as turning base to final in the landing configuration, the aircraft is flown uncoordinated (for example hurried in the turn with excess rudder, causing a banked skid) and enters a stall/spin as a result. The key to being safe from inadvertent spins is not as much in being able to recover from the spin, but in being able to recognize the symptoms and taking corrective action before the spin occurs.

Recovery from the spin is accomplished by rudder opposite to the direction of yaw to stop the yaw, thereby equalizing the drag on the two wings and stopping the rotation, followed by forward stick (down elevator) to lower the angle of attack and unstall the wing. Under no circumstances should aileron opposite the bank be applied. It is the novice's instinctive reaction to pick up the dropped wing with aileron and it is entirely wrong. *Opposite aileron makes the spin worse*; because of the rotation, the inside (fully stalled) wing is actually encountering some airflow from behind. When opposite aileron is applied, the inner wing's camber increases, creating some lift, which causes the wing to rise, flattening the spin; however, due to the continuing yaw, the aircraft remains firmly established in a spin. The only place for the aileron in the spin is in neutral.

A good exercise to cure aileronitis is to stall the airplane, keep it in the stall, and keep the wings straight and level by applying opposite rudder to recover (pick up) a dropping wing (this exercise is sometimes referred to as the *falling leaf*).

Competition spins differ from training spins and inadvertent spins in two important respects:

- The aircraft is yawed 3–5 knots above the published stall speed to prevent the aircraft from yawing in the wrong direction
- On recovery, a distinct vertical downline is established before pull out into straight and level

Let's see how the upright spin is flown.

FLYING IT

Except for the application of rudder to induce the yaw, setting up the spin is quite similar to setting up a power-off stall. You will lose more altitude in the spin than in most other maneuvers you have learned to date, so be extra conservative and fly high with plenty of altitude. Start the spin at least at 3,500 feet agl in a typical aerobatic training machine such as the Decathlon (consult your aircraft manual for specific altitude requirements and recovery procedures).

1. Line up over a prominent straightline landmark for orientation and establish straight and level flight. Smoothly retard the throttle to idle and maintain altitude by increasing pitch; gradually come back on the stick (up elevator) as you retard the throttle. The objective is to get the nose as high as possible without gaining any altitude.

2. At an airspeed 3–5 kts faster than the published stall speed, introduce yaw in the desired direction of rotation by smoothly but briskly applying full rudder in that direction.

3. As the aircraft starts to yaw, bring the stick smoothly and briskly full aft to induce the stall, and hang on for the ride. Because of the yaw, the wing inside the direction of yaw slows down, the wing on the outside speeds up; the inside wing stalls fully and drops; the outside wing starts rotating around it, and you are in the spin.

4. Maintain full in-yaw rudder and full aft (up) elevator. Look directly over the nose straight at the ground. Looking elsewhere, such as at the horizon, might be very disorienting. Count out each half rotation to maintain orientation. If you don't count them, you might go through more rotations than you had planned (and lose proportionally more altitude). For a one-and-a-half turn spin, you would count "half . . . one . . . "; you would not count the next value "one and a half" because upon reaching the one-and-a-half turn mark you have to be already recovered from the spin.

5. With a ½ to ¼ turn to go to the point where you want to recover (in this case the 1 and a ½ turn mark), initiate recovery:
    Smoothly apply full opposite rudder to stop the rotation.
    When the rotation is almost stopped, apply forward stick (forward of neutral), to unstall the wings.

6. As you reach the desired recovery heading (confirmed by the ground reference line), neutralize all controls. The aircraft will then be in a dive, recovered from the spin.

7. If the spin is a competition spin, establish a vertical downline and then smoothly pull out of the dive, restoring power. If the spin is not a competition maneuver, pull out as soon as the aircraft is recovered from the spin and established in the dive.

8. (Inadvertent spin recovery.) If you entered the spin inadvertently, the spin recovery technique is the same as described, but because the aircraft will most likely be developing full power in the spin, the first thing to do is *retard power to idle*.

Many pilots forget to cut the power before attempting to recover because they are used to having the power off well before they deliberately enter a spin and reach the recovery stage. Many aircraft types will not recover with the power on, so develop the "retard power" command into a reflex reaction to an inadvertent spin.

Power retardation should be followed by the standard recovery method of opposite rudder, followed by stick forward as the rotation is about to stop, neutral controls when it does stop, and a pull out from the dive.

COMMON ERRORS

There are several common errors flying the competition spin, some with safety implications.

Rudder is not applied early enough in initiating the spin. If rudder is applied late during spin entry, the aircraft will take its time beginning to rotate (the yaw input is more effective at the faster speed), and excessive altitude will be lost in the spin as the aircraft "settles" before beginning to spin.

The pilot fails to count the rotations during the spin. This error leads to disorientation. The rotation rate in the spin is quite high and when orientation is lost, it is difficult to reestablish. Typically, novice pilots do quite well recovering from the spin, but often go through more rotations than they think they did.

The rudder is not neutralized prior to pull out. This error does not cause problems in typical training aircraft, but can cause a spin in the opposite direction in high-performance competition machines.

The pilot slightly eases off the rudder and elevator during the spin. In low-performance trainers, this error can cause the aircraft to transition from the spin into a spiral dive, which the student confuses with the spin. One big difference is obvious in the cockpit during the spiral dive; because the aircraft is not stalled, the airspeed increases alarmingly.

BEGGS/MÜLLER EMERGENCY SPIN RECOVERY TECHNIQUE

Pilots become confused in spins from time to time and are unable to recover. Later they describe their experience and insist that they tried every recovery technique they knew, but nothing worked. Over the years, pilots jumped out in such situations and many noticed an interesting fact: As soon as they let go of the controls to unstrap themselves when they prepared to jump, the aircraft recovered from the spin on its own accord. Others watched with a sinking feeling hanging from their parachute straps as their pilotless aircraft recovered and flew off to an inevitable crash.

Gene Beggs in America and Eric Müller in Switzerland decided to experiment systematically with letting the spinning aircraft do what it will and found a recovery method that works with many aircraft types. Experiment at altitude to see if it works for your aircraft.

First attempt to recover conventionally from a spin, whether intentional or unintentional. If you experience no response, apply the Beggs/Müller emergency recovery technique:

- *Power off*
- Let go of the stick
- Rudder opposite the yaw
- When the aircraft has stopped spinning, neutralize controls and pull out of the dive normally

INADVERTENT INVERTED SPINS

From the perspective of the aircraft, the inverted spin is not different from the upright spin; however, from the perspective of the pilot and the control inputs required to spin inverted there is a difference because the pilot and the controls are upside down in mirror image to being right side up. If you understand inverted flight and feel comfortable flying inverted (including turns), you should have no problem understanding the inverted spin (for a refresher see the chapter on inverted flight).

To get the nose up so that the aircraft would spin in the inverted position, the stick has to be pushed forward instead of being brought aft.

To induce the yaw required to enter the inverted spin, you have to step on the rudder opposite the intended direction of the rotation (just like the inverted turn, remember?) *A major difference from the upright spin*: In the inverted spin, the direction of yaw is opposite to the direction of rotation.

To recover from the inverted spin, the control inputs have to be reversed: rudder opposite the yaw is applied to stop the rotation; as the rotation slows, aft stick is applied to unstall the wing. When the aircraft is no longer spinning, the controls are neutralized, and the pilot pulls out of the dive.

Rudder opposite the yaw in the inverted spin is rudder in the direction of the rotation. It is more difficult to establish the direction of rotation inverted because of a higher rotation rate; therefore, concentrate on "opposite the yaw." Simply put, push the rudder opposite to the rudder that is applied in the spin. If you are confused about whether or not you pushed the correct rudder, check the rotation rate. If rotation does not noticeably slow down after you pushed opposite rudder, you know you pushed the wrong one. Don't panic, push the other one, wait for the rotation to slow, and come aft with the stick.

The Beggs/Müller recovery method also works for the inverted spin. Just remember to apply rudder opposite the yaw:

- *Power off*
- Let go of the stick
- Rudder opposite the yaw
- When the aircraft has stopped spinning, neutralize controls and pull out of the dive normally.

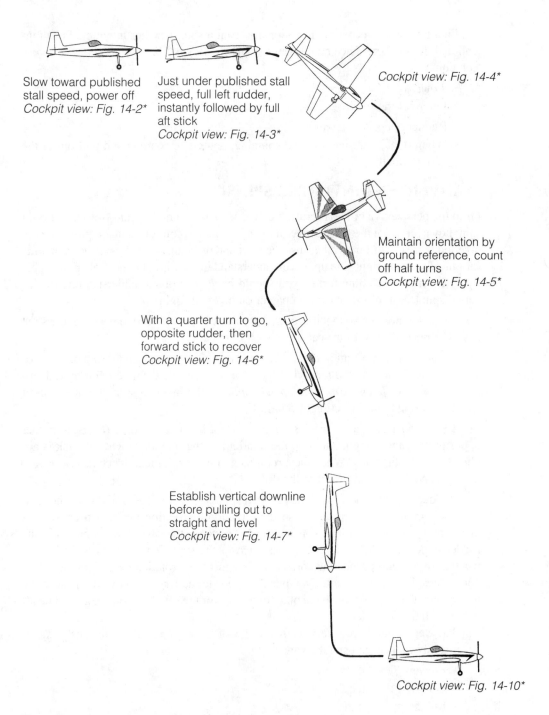

Slow toward published stall speed, power off
Cockpit view: Fig. 14-2*

Just under published stall speed, full left rudder, instantly followed by full aft stick
Cockpit view: Fig. 14-3*

Cockpit view: Fig. 14-4*

Maintain orientation by ground reference, count off half turns
Cockpit view: Fig. 14-5*

With a quarter turn to go, opposite rudder, then forward stick to recover
Cockpit view: Fig. 14-6*

Establish vertical downline before pulling out to straight and level
Cockpit view: Fig. 14-7*

Cockpit view: Fig. 14-10*

Fig. 14-1. *One-turn competition spin to the left. (*Photo sequence is of a one-turn spin to the right.)*

Fig. 14-2. *Photo sequence of one-turn spin to the right. Establish straight and level, power off, keep applying aft stick as aircraft slows.*

Fig. 14-3. *Full right rudder 3–5 knots above published stall speed, instantly followed by full aft stick.*

Fig. 14-4. *Going through first quarter turn.*

Fig. 14-5. *Count off one-half a turn.*

Fig. 14-6. *Quarter turn to go, opposite rudder, stick full forward.*

Fig. 14-7. *Neutralize controls, establish vertical downline.*

Fig. 14-8. *Begin pullout.*

Fig. 14-9. *Continue pullout.*

Fig. 14-10. *Exit the spin straight and level.*

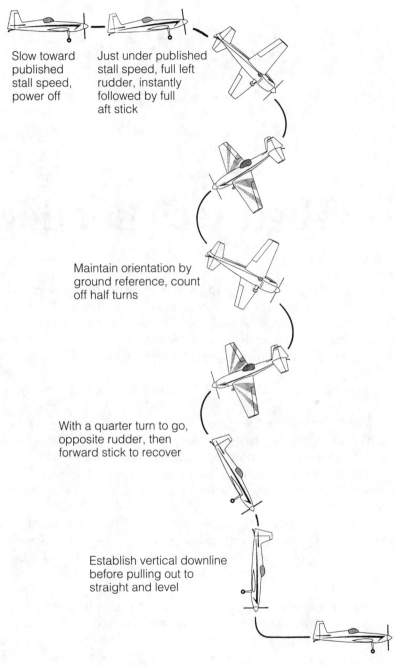

Slow toward published stall speed, power off

Just under published stall speed, full left rudder, instantly followed by full aft stick

Maintain orientation by ground reference, count off half turns

With a quarter turn to go, opposite rudder, then forward stick to recover

Establish vertical downline before pulling out to straight and level

Fig. 14-11. *Two-turn competition spin to the left.*

High-tech hot dogs

Sport aerobatics is the world of the roll, the loop, the hammerhead, and the spin. Eavesdrop on a group of jet fighter pilots and you will hear of aspect angles, the high speed yo yo, the straight ahead pop up, and the 135° slice. How different is the ultimate winner-take-all aerobatic competition?

The fundamentals are the same. The matter of physics is certainly the same, and if anything, fighter pilots are even more concerned with energy management than aerobatic pilots. "High, slow; low, fast" is so ingrained in their mind that it could be their motto.

The differences are in the objective of the aerobatic maneuvers they fly, and the aircraft performance they have to accomplish it. While the aerobatic pilot wants to fly the perfect maneuver every time, the fighter pilot's sole objective is to nail the other guy first. The aerobatic pilot follows the sequence card of strictly defined maneuvers with great precision. The fighter pilot does anything it takes to "maneuver to advantage."

Interestingly enough, fighter pilots routinely pull less G force than advanced aerobatic pilots, for sound tactical reasons. A 9G pull in a fighter as maneuverable as the F-16 confines its movement to such a narrow area that it could be lethally sprayed by the opponent's guns. A fighter must keep streaking across vast spaces quickly to stay alive.

Fighters can and do maintain high G loads for much longer than any sport aerobatic airplane. A sustained 6G turn is common and the risk of G Loc is ever-present. An F-16 can *accelerate* in a 6G turn. The fighter's immense power gives a whole new meaning to proportions. A 5G loop takes about 10,000 feet and has to be started at 450 knots. An F-15 can accelerate to 500 knots, reduce power to idle and do a loop. The *floor* of air-to-air combat exercises is typically 10,000 feet.

The maneuvers, however, as specialized as they are, have their roots firmly in the world of basic aerobatics. Consider a typical maneuver from each of the two branches of air-combat maneuvering, air-to-air and air-to-ground combat.

In air-to-air combat, the fighter pilot maneuvers for a position of advantage against another aircraft. The ground is there in the background as an element to be avoided, but the primary reference point is the other aircraft. The pilot strives to achieve the aspect angle that would ultimately place the adversary in the fighter's sights long enough to kill it.

Place yourself in the cockpit of an F-15. Your adversary is to your right, at the same altitude, turning hard, aware that from your aspect angle at your high speed, you can't turn inside him and get into firing position. But you have an option: pull hard on the stick, soaring above him, converting your kinetic (airspeed) energy into potential (altitude) energy; as he desperately keeps turning, pull over in a looping flight path, simultaneously rolling to the left to the inside of his flight path and ruddering the nose down to move him into your sights. When you start heading down, your potential (altitude) energy rapidly builds into kinetic (airspeed) energy, and if you timed it right you'll just have to fire.

What you did is called the *high-speed yo-yo*, but it is really just a combination of a loop, a roll, and a chandelle for a special purpose.

In air-to-ground combat, the fighter maneuvers against the ground, attempting to destroy a target that is presumably defended. The greatest position of advantage is accomplished by remaining undetected as long as possible and hitting the target in the least amount of time over it. A favored technique to accomplish this position of advantage is the *straight-ahead pop-up*. You're flat on the deck, roaring toward your target at 450–500 knots. You pull up at 5–6Gs at a predetermined point just before the target, roll inverted on the upline, look down, acquire the target, pull down toward it at the apex altitude of your climb, roll upright on the downline keeping the target stationary in your sights, blast it to kingdom come, and jink the heck out of there.

It is called the pop-up, but what you really flew were classic rolls along uplines and downlines. Times and altitudes vary, but a typical pop-up would get a fighter at 450 knots up to its apex altitude at a phenomenal climb rate of 10,000 feet per minute and keep it exposed over the target for a mere 10 seconds.

To be practiced safely and understood fully, mock air-combat must be learned from appropriately qualified instructors, like any other form of aerobatics. Fortunately for civilians, an increasing number of civilian air-combat schools are being established by ex-military instructors flying military trainers such as the Beech T-34, the Siai Marchetti SF .260, and even the venerable T-6 Texan. Check out the school's credentials carefully, load the gun-video cameras and ask them to show you a 135° slice.

15
Advanced maneuvers

MANY OF US ARE CONTENT WITH CONSTANTLY FINE-TUNING OUR fundamental aerobatic skills when we have mastered the basic maneuvers. But for the more persistent and adventurous among us, there will come a time when basic aerobatics will become as routine as a VFR hop around the neighborhood. We will then be eager for a new challenge and will readily find it in the world of advanced aerobatics.

While the point beyond which aerobatics is called "advanced" is somewhat subjective, greater degrees of complexity, less margin for error, and greater demands on aircraft performance and the human body distinguish advanced maneuvers from the basics. The world of competition aerobatics recognizes this distinction in the types of maneuvers assigned to the various categories of competition. In the aerobatic community, maneuvers that are generally assigned only beyond Sportsman level competition are often referred to collectively as advanced aerobatics and this is the definition we use here. (A gray area where we err on the "advanced" side is the upright snap roll, which has been assigned at times in the Sportsman sequence.)

Advanced maneuvers are beyond the instructional scope of this book; however, they are introduced because advanced aerobatics is the sport at its most exciting level, demanding a degree of competence well worth aspiring for under the guidance of a qualified advanced aerobatic instructor.

SNAP ROLLS

The snap roll is an autorotational maneuver. It can be thought of as a spin along the direction of flight in which the aircraft was traveling prior to beginning the maneuver

(FIG. 15-1). This is made possible by such a brisk rate of change in pitch and yaw that the direction of the relative wind is unable to keep pace and remains unchanged. The abrupt control movements imply a high acceleration rate (Gs); therefore, entry speed control is critical and the maneuver must be performed well below maneuvering speed to allow for a good safety margin.

The snap roll is performed by establishing the proper entry speed and briskly applying full rudder in the desired direction of rotation followed in a fraction of a second by aft stick to create an accelerated stall. These control movements are not separated but "jointed," with a split-second delay in stick application. The yaw caused by the rudder slows the inside wing (the one on the side of the applied rudder) and accelerates the outside wing. When the stick is briskly brought aft to induce the accelerated stall, the inside wing stalls and the outside wing rotates around the inside wing as if in a spin. Slightly easing the stick forward from the aft position when rotation has begun increases the rate of rotation because it decreases the drag caused by full aft elevator.

Recovery is accomplished by full opposite rudder and stick forward of neutral (as from a spin). The objective is to time the recovery to bring out the aircraft into straight-line flight on the same heading and altitude at which it was flying at the beginning of the maneuver.

The snap roll uses up a considerable amount of energy as can be seen from the high G meter reading and the big drop in airspeed by the end of the maneuver. The need to prevent an excessive loss of speed due to the high drag caused by the horizontal stabilizer is achieved by slight forward stick when rotation has begun. Because the snap roll is started at a slow airspeed to begin with, the snap roll can get the pilot into trouble if mishandled, most commonly resulting in an inadvertent normal spin.

The snap roll is a very common maneuver in competition flying. It is frequently performed in the horizontal as well as the vertical plane. A popular variation is several snap rolls performed in sequence.

OUTSIDE SNAP ROLLS

The outside snap roll is performed exactly as the inside snap roll except forward elevator is applied briskly (instead of aft elevator). In essence, it is an inverted spin done along the direction of flight in which the aircraft was moving when the maneuver was initiated and along which it continues to move. The change in pitch and yaw is too brisk for the relative wind to keep pace.

When rotation has begun, the stick is eased slightly aft of forward to reduce drag caused by the horizontal stabilizer and thus accelerate the rotation. Recovery is accomplished by applying full opposite rudder and neutral stick. The objective is to time the recovery to bring out the aircraft into straightline flight on the same heading and altitude at which it was flying at the beginning of the maneuver.

The negative snap roll is frequently seen in competition. It is an extremely uncomfortable maneuver because of the high negative G caused by the brisk control movements.

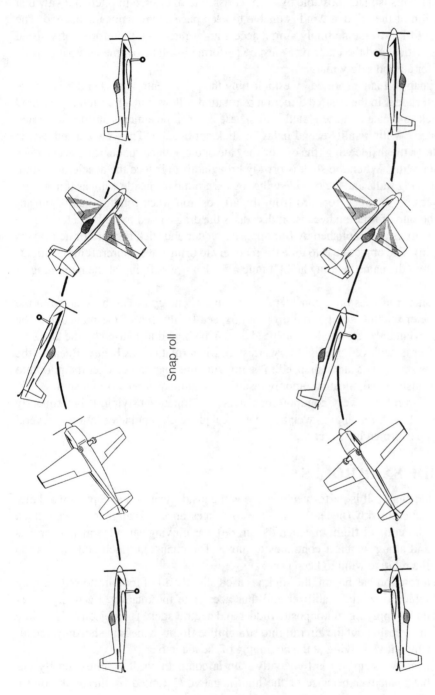

Snap roll

Inverted snap roll

Fig. 15-1. *Snap rolls.*

TAILSLIDES

In a tailslide, the aircraft is pulled into vertical straightline flight and is allowed to completely run out of energy, at which point it comes to a momentary standstill and then slides tail first back toward the ground. Eventually, it pivots around its center of gravity and the nose swings down toward the ground (FIG. 15-2).

If the aircraft is held perfectly vertical on the way up, the slide down tail first will be prolonged and speed might build up so much in this position that the tail surfaces might be damaged by the reverse airflow; therefore, the proper technique is to hold the aircraft on the upline either slightly nose forward (positive) or nose aft (negative) of the vertical upline to initiate the swing around the center of gravity soon after the beginning of the slide backward.

From a positive upline, the aircraft should pivot forward (wheels down) to recover and can be aided in doing so by applying full aft stick. This exposes the maximum horizontal tail surface area to the reverse airflow, which pushes the tail up (there is practically no lift generated by the horizontal tail surface). From a negative upline, the aircraft should pivot over on its back to recover (wheels up) and can be aided by full forward stick exposing maximum horizontal tail surface to the reverse airflow, which pushes the tail up.

Many aerobatic aircraft are not approved for the tailslide because of potential damage to the tail surfaces.

OUTSIDE LOOPS

The outside loop is a 360° circle in the vertical plane similar to the inside loop except that the top of the airplane is facing the outside of the circle formed by the loop. The maneuver is accomplished by pushing the stick forward to fly the aircraft around in a circular pattern (FIG. 15-3). The pilot, whose head points away from the center of the circle, experiences high negative G during the outside loop, which physically is one of the most uncomfortable aerobatic maneuvers.

The outside loop is flown in one of two ways. It can be entered from upright straight and level flight at a slow airspeed, from which the aircraft is progressively pushed into the maneuver downward; or the aircraft can be rolled inverted at a fast airspeed and pushed into the maneuver upward.

INVERTED SPINS

The inverted spin is identical in nature to the upright spin except that the aircraft is stalled upside down, in a negative, or stick forward, condition. The big difference in the inverted spin in comparison to the upright spin is that inverted the yaw and roll (rotation) are in opposite directions, which causes a spinning motion that can be most disorienting to the beginner. The important thing to remember is that to stop the rotation the yaw has to be neutralized, which is accomplished by applying rudder opposite the *yaw* (*not* opposite the direction of rotation, a critical distinction).

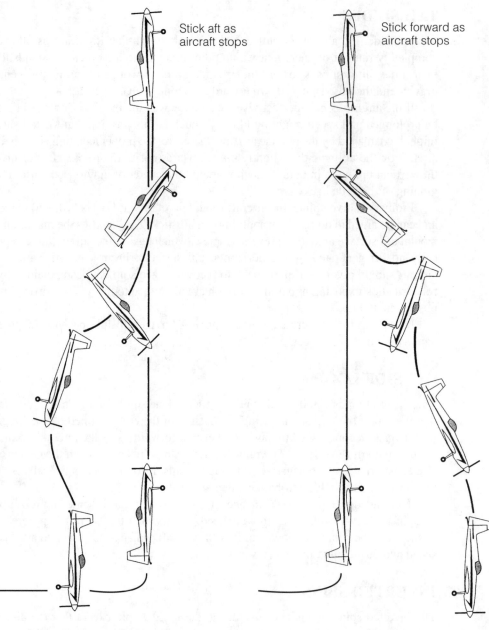

Stick aft as
aircraft stops

Stick forward as
aircraft stops

Stick-aft tailslide

Stick-forward tailslide

Fig. 15-2. *Tailslides.*

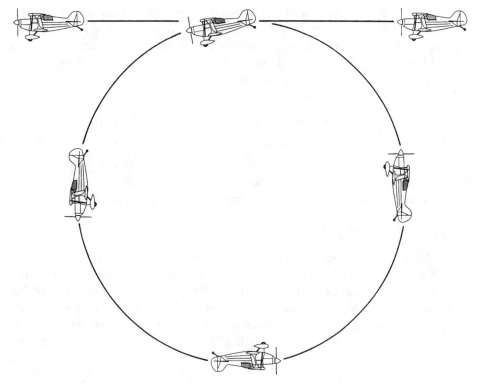

Fig. 15-3. *The outside loop.*

The inverted spin is initiated from the inverted position by applying rudder in the desired direction of yaw and forward stick to stall the wings. The control inputs are applied approximately 3–5 knots prior to stall to assure a positive break in the desired direction of the spin.

Recovery is accomplished by applying rudder opposite to the *yaw* to stop the rotation, followed by aft stick to unstall the wings. The inverted spin will rotate faster than an upright spin (due to higher autorotative forces), but will also recover at a faster rate.

If you are confused, do not be discouraged. A whole book could be written on inverted spins alone. The inverted spin is initially extremely disorienting and can be quite confusing; thorough dual instruction from a qualified instructor is absolutely the only way to learn the inverted spin. The inverted spin is a staple maneuver of the unlimited aerobatic competitor.

VERTICAL ROLLS

During a vertical roll the aircraft is rolled around its longitudinal axis on a vertical upline or downline. Rolls might be a full 360°, or they might be half rolls, quarter rolls, or a variety of hesitation rolls.

The vertical roll is accomplished with ailerons. The aircraft has to be established in a perfect vertical upline or downline before the roll is started. The pilot looks at a wingtip in relation to the horizon to monitor the accuracy of the maneuver and stops the roll based upon visual reference points noted toward the horizon. A wingtip displacement means that the vertical line has been compromised and is generally controlled with minor elevator movements:

- Forward stick to correct a wingtip rising relative to the horizon
- Aft stick to correct one sinking relative to the horizon

A lot of energy is required to establish a vertical upline long enough for performing rolls, so the maneuver must be started from a high entry speed.

Vertical rolls are popular elements of competition sequences and airshow displays.

KNIFE-EDGE FLIGHT

During knife-edge flight, the aircraft is banked 90° and is flown in a straight line along a horizontal flight path. Knife-edge flight always excites spectators who are not well versed in aerodynamics because the aircraft seems to fly without any apparent source of lift. The source of lift that makes knife-edge flight possible is the air flowing over the fuselage (FIG. 15-4). The fuselage is generally symmetrical and streamlined in a shape that resembles an airfoil. So, while it is not nearly as efficient as a wing, it is certainly a source of lift.

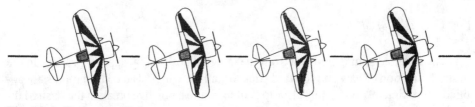

Fig. 15-4. *Knife-edge flight.*

Because the fuselage is not an efficient source of lift and it is symmetrical, when the aircraft is rolled into the knife-edge position, the fuselage's angle of attack has to be substantially increased to generate lift sufficient to maintain horizontal straightline flight (FIG. 15-5). This is accomplished by top rudder because of the reversed roles of elevator and rudder controls in the 90° bank. Because the lift generated by the fuselage increases with an increase in the speed of the airflow over it, the maneuver is best started from a fast entry speed.

Not all aircraft have good knife-edge capabilities. The fat rotund fuselages of some biplanes, among them the Pitts, seem to have the best airfoil characteristics.

HESITATION ROLLS

Hesitation rolls are a variation of the slow roll in which the aircraft is momentarily stopped at various points during the roll: 3-point, 4-point, 8-point, and 16-point. The

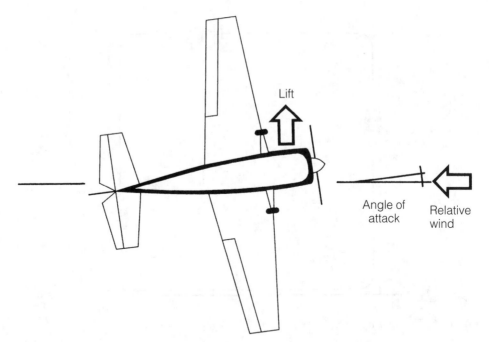

Fig. 15-5. *Knife-edge flight makes the fuselage act as an airfoil, generating lift.*

roll segments are equal during hesitation rolls; thus, during the 4-point roll the aircraft is stopped every 90° and during the 8-point roll it is stopped every 45°. It is easy to see why a pilot has to be able to fly a perfect slow roll every time before attempting a hesitation roll.

The hesitation roll is flown as if a slow roll, except at the precise points where the aircraft is to be momentarily stopped, the aileron is brought back to neutral for an instant. It takes great coordination and a steady hand to get the points correct and prevent them from bobbling. Because of the roll interruptions, the hesitation roll takes longer than the slow roll, requiring a faster entry speed in such high drag aircraft as some antique biplanes.

SQUARE LOOPS

During the square loop the aircraft flies a square path in the vertical plane. Each corner of the square is formed by a quarter loop.

Though the figure forms a perfect square (FIG. 15-6), the technique required to fly each segment varies slightly. A great amount of speed is needed to fly a good vertical upline. During the top horizontal segment, the pilot has to be precise in inverted flight and has to accurately judge the length of the segment to equal the vertical upline. The vertical downline segment has to be flown with partial power to prevent excessive speed buildup in anticipation of the final high-G pullout to complete the square loop.

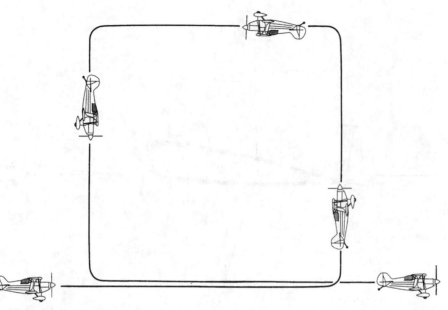

Fig. 15-6. *The square loop.*

More challenging variations of the square loop include a diamond-shaped loop in which each side is a 45° downline or upline, an 8-sided loop, and inverted square loops.

ROLLING CIRCLES

Rolling circles are one of the most challenging coordination maneuvers. During the rolling circle, the aircraft is flown in a circle in the horizontal plane and at the same time the aircraft is rolled around its longitudinal axis. Perhaps one full roll during the circle (FIG. 15-7), a full roll in each half of the circle, or a full roll in each quarter of the circle.

The trick to the rolling circle is to initiate a turn and simultaneously establish a roll rate that will result in a completed roll at the appropriate point on the circle depending upon the number of rolls planned for the circle. Consider, for example, a *4-roll inside rolling circle* to the left; "inside" means that the roll is in the direction of the turn. Rolls in the rolling circle may also be performed to the outside of the circle, away from the direction of the turn.

The first step after establishing maneuver entry speed is to select two points on the horizon: one point that is halfway between straight ahead and the left wingtip (the *45° point*) and one point on the wingtip (the *90° point*). When this latter point is reached, one quarter of the circle will be completed and the aircraft will have had to complete a full roll upon reaching this spot.

Left rudder is added to start the turn and the roll is commenced at the same time. As vertical bank is reached, up elevator (back stick) keeps the turn going and top rud-

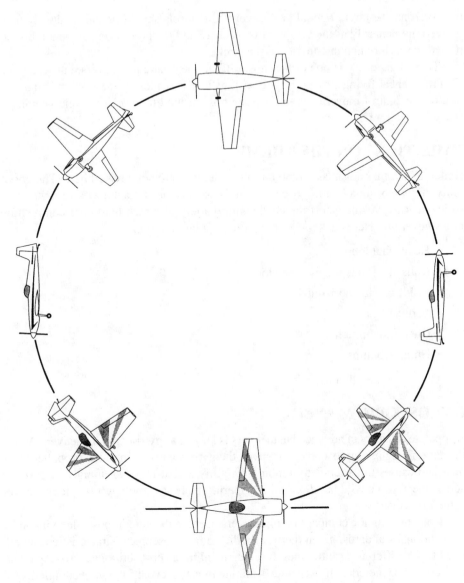

Fig. 15-7. *The rolling circle—one of the most difficult aerobatic figures—is an inside rolling circle to the left with a single roll in the 360° turn (viewed from above).*

der keeps the nose on the horizon. When the aircraft is inverted, it should be at the 45° point halfway toward completing the first quarter of the circle. Right rudder now keeps the turn going and down elevator (forward stick) keeps the nose on the horizon. The pilot has to look out the right side to pick up the approaching 90° point. When vertical bank is reached, forward stick keeps the turn going and left rudder keeps the nose on the horizon. The aircraft should roll level and controls should be harmoniously neutralized momentarily as the 90° point is reached, and the process is then smoothly repeated for each remaining quarter of the circle.

The rolling circle is one of the most difficult aerobatic maneuvers, but when you are proficient at flying slow rolls, you will instinctively pick up the rhythm and direction of the control movements to keep the nose around the horizon while keeping it around in a circle.

VARIATIONS ON THE THEME

Hundreds of other advanced combination maneuvers are not covered here. The possibility of composing a seemingly endless series of advanced maneuvers from the aerobatic catalog gives the sport the challenging variety of which hard-core competition pilots never tire. Here are some of the more popular variations:

- Hexagonal loops
- Rolls at various stages of the loop
- Inside outside horizontal eights
- Vertical eights
- Outside hammerheads
- Humpty bumps

The sky is literally the limit.

GYROSCOPIC MANEUVERS

A separate category of advanced maneuvers is known as gyroscopic maneuvers. When the aircraft begins a gyroscopic maneuver, the gyroscopic action of the propeller takes over and causes the aircraft to tumble gracefully and at times seemingly impossibly across the sky in a wild but discernible pattern. The best known gyroscopic maneuver is the Lomcovak.

None of them are competition maneuvers because they are impossible to judge by conventional standards, given the widely differing gyroscopic characteristics and capabilities of different aircraft types. Instead, in unlimited competitions, gyroscopic maneuvers can be included in a separate 4-minute freestyle event following completion of the regular competition. The freestyle is judged entirely separately by different standards; it is great fun and the opportunity for unlimited competitors to engage in a bit of airshow-style flying.

Because of their limited role in competition flying and their complexity, a more detailed discussion of gyroscopic maneuvers is beyond the scope of this book.

16
Developing aerobatic sequences

LEARNING TO FLY INDIVIDUAL MANEUVERS IS ONLY THE FIRST STEP FOR the aspiring aerobatic pilot, starting the process of acquiring the tools of the trade. Aerobatics becomes the exhilarating experience it is meant to be when its tools are put to work in unison, when the maneuvers are strung together in a sequence to suit your mood, to demonstrate your competence to a group of judges at a competition, or to please the crowds.

Certain issues of energy management must be taken into account when composing aerobatic sequences, and the capabilities and limits of the airplane and the pilot must also be considered. But sequence composition is also an art, almost like composing music. The maneuvers are the notes. How they are arranged determines if the composition will be lyrical and harmonious, or discordant and jarring.

In aerobatic competitions, pilots are given sequences to perform (compulsory flights), or fly sequences of their own design (freestyle flights). The ultimate purpose of this book is to get the aerobatic student to the point of being able to enter competitions at the beginner level. This chapter introduces the fundamentals of sequence composition by preparing a sequence for the *Sportsman Category*, which is the level that most fledgling aerobatic pilots enter for their first competition. Flying the sequence at altitude is discussed. (Flying in the "aerobatic box" is discussed in chapter 18.)

By the time you work your way through this chapter with your instructor, you will be ready to safely enjoy recreational aerobatics at your leisure, confident in the knowledge that you have been trained to entry-level competition standards. Or, you can move on into chapter 18 and the enticing world of aerobatic competition, where you will learn all about the intricacies of the aerobatic box (and how to bring your maneuvers into it), using aerobatic shorthand, and how to prepare for and fly your first aerobatic competition.

AEROBATIC SEQUENCE FLYING IS ENERGY MANAGEMENT

We have said it before and we say it again: aerobatics is energy management. Nowhere is this more obvious than in flying an aerobatic sequence. When you fly an individual maneuver, you take the time to first position the aircraft at the right altitude and then you build up the right airspeed to start the maneuver.

When flying maneuvers sequentially, you do not have that luxury. When you complete one maneuver, you have to be at the right altitude and airspeed to fly the next maneuver, which you have to complete at the right altitude and airspeed sufficient for the following maneuver, and so on until the end of the sequence. This requirement leads us to two important points regarding the sequence in which aerobatic maneuvers can be flown:

- To begin a high-energy maneuver, you have to be at a high airspeed. A high airspeed implies that your previous maneuver was also a high-energy maneuver, or that you dove to attain the high airspeed required for your upcoming maneuver. If this maneuver is one that ends at the same altitude where it started (such as a loop or half Cuban eight), it will also end at about the same high airspeed as the entry speed.

- To maintain the available energy represented by the high airspeed, your next maneuver has to be another high-energy maneuver that also ends at the same airspeed and altitude, or ends at a low airspeed, but at an altitude equivalent to the highest altitude reached in the previous maneuver (so that you can dive to build energy). This technique keeps the amount of energy available approximately constant throughout the sequence and gives you maximum flexibility in the maneuvers that you can fly.

Consider your position if you squander the high exit airspeed by doing a drawn out slow roll instead of trading it for altitude, and at some point in the sequence converting it back into airspeed again. You would not have sufficient airspeed for another high-energy maneuver. To reacquire the energy you squandered in the roll, you could dive to an altitude that is lower than the base of your last high-energy maneuver, but you can do that only so many times before you run out of altitude. Eventually you would be restricted to maneuvers that the aircraft is capable of doing from its maximum straight-and-level speed or you would have to break your sequence, climb to altitude, and start over.

The airspeed at which you finish a maneuver restricts your choice of the next maneuver to those that have an entry speed equivalent to your exit speed. On exiting a loop, you would have to keep on trucking in straight and level flight for a long time if you wanted to do a snap roll or a spin. An Immelmann or a hammerhead, however, could be entered right away, so it would be a good choice.

You can see from the energy management requirements of an aerobatic sequence that you should plan to fly low-energy horizontal maneuvers, such as a slow roll or a snap roll, near the ceiling of your chosen height band, and you should establish the floor of your height band at an altitude to which you have to dive to build up the required airspeed for your high-energy vertical maneuvers. If you want to follow a high-energy maneuver with a low-energy maneuver, your high-energy maneuver should end at a high altitude at a slow airspeed. From this position, you can fly your low-energy maneuver and then dive for another high energy maneuver. For example, enter a slow roll or a snap roll from a suitable high-energy maneuver, such as a half loop or an Immelmann.

The extent to which you can sequence various types of maneuvers is also influenced by two limitations:

- Aircraft performance limitations. High-performance competition aircraft are able to fly certain maneuvers in sequences that a training aircraft might not be able to match.
- Human limitations. Respect the limits of your aerobatic skills and stamina. Don't overload a sequence with a long string of complicated maneuvers that are pushing the edge of your abilities and endurance. Flying maneuvers in sequence is physically more demanding than flying them individually. After a challenging maneuver, there is no time to take a breather to settle down and prepare for the next one. Know your limits and if you discover in flight that you have overdone the workload, stop the sequence immediately. Pushing your endurance not only makes aerobatics an ordeal, but also downright dangerous.

COMPOSING AN AEROBATIC SEQUENCE

The quickest way to understand the issues and techniques of composing aerobatic sequences is to walk your way through a simple example. Consider this typical Sportsman Category sequence (represented in FIG. 16-1 at the end of this chapter), flown in a Decathlon:

1. Loop
2. Immelmann
3. One-and-a-half-turn spin
4. Hammerhead
5. Reverse half Cuban eight
6. Half loop
7. Slow roll

In the interest of safety, you will fly your sequence at altitude; pick a floor altitude below which you will not descend—3,000 feet agl is a good floor altitude. Your first maneuver will be a loop, a high-energy vertical maneuver and a good opening element.

In the Decathlon, you will enter the loop at about 140 mph. The question is, at what altitude should you start the loop if you want the floor of your height band to be 3,000 feet? Think ahead to the rest of the maneuvers that you want to fly in your sequence.

The maneuver that will require the most energy is the reverse half Cuban eight. Its entry speed is about 160 mph in the Decathlon. So, if you started the loop at 3,000 feet from 140 mph, you would have to dive below 3,000 feet to attain the energy represented by the reverse half Cuban eight's higher entry speed. Also, you will not execute every maneuver to 100 percent perfection, resulting in some energy loss, and the aircraft's drag will nibble away at the altitude you can regain from maneuver to maneuver; thus, to give yourself room for the reverse half Cuban eight and some margin of error, you had best start the loop at 3,500 feet.

When you complete the loop, the aircraft will be at a fast airspeed and a low altitude relative to the floor of your selected height band. This situation dictates another high-energy maneuver, but which one to choose if you want to get into some slow speed maneuvers soon? An Immelmann fits the bill. It converts the airspeed into altitude and slows the aircraft to an appropriately slow speed for your next maneuver.

When you roll upright from the Immelmann, you will be at a high altitude and near the published stall speed, which is the ideal position for a spin. A nice feature of the spin is that you can come out of it in any direction, so it is well-suited for reversing the direction of your flight path. A one-and-a-half-turn spin will reverse your direction and as you exit from the spin, you can convert altitude to airspeed for another high-energy maneuver.

You might be somewhat below 3,500 feet at this stage by the time you build the necessary airspeed for your next maneuver, the hammerhead, but you should be well within your 3,000-foot floor altitude. The hammerhead is another useful direction reversal maneuver. As you recover from the hammerhead, you will be going opposite your direction of entry, and ideally set up for the next maneuver, the reverse half Cuban eight.

The reverse half Cuban eight requires the greatest amount of energy in this sequence, the fastest entry speed (about 160 mph). The altitude at which you attain 160 mph is a good indicator of how efficiently you have been managing your energy so far. If you have to descend below 3,000 feet, your self-imposed floor, you haven't been converting airspeed into altitude and vice versa at the optimal rate. Your maneuvers are probably a bit sloppy.

The reverse half Cuban eight ends at high speed near the floor of your height band, so you are ready for another high-energy maneuver. You might be getting tired and would like to wrap up your sequence with a less strenuous, low-energy maneuver, so you will want to slow down. Converting airspeed into altitude through a half loop will slow you down nicely, and put you in a perfect position to do a leisurely slow roll and call it a day.

LEARNING TO FLY AN AEROBATIC SEQUENCE

When you and your instructor have come up with a basic aerobatic sequence and you feel comfortable with the composition techniques and the reasons behind them, it is time for the two of you to go up there and fly your creation. Initially you might want to break down the sequence. Become comfortable flying three or four elements at a time before you fly the whole sequence from beginning to end. If you have any difficulty with a particular maneuver, practice it on its own until you get it right, before doing it again in a sequence. When your instructor thinks you are ready, practice the sequence on your own.

You will find that as you become more and more proficient in flying sequences, you will be thinking less and less of each individual maneuver. Your control inputs will become as natural and automatic as in a standard-rate turn. You will quickly anticipate minor mistakes as they are about to happen, and you will correct them almost instinctively.

Aerobatics will really start to be fun!

You will start to notice the subtleties of your airplane's performance in different weather conditions. You will learn to expect less from an airplane on hot and humid days, compared to dry, cold days. And if you experiment with the same sequence in different aircraft types—*always with an experienced instructor, first*—you will notice the sometimes substantial differences in capabilities between different aircraft types that might require modifications to your sequence.

If your interest is recreational aerobatics, you are ready to enjoy yourself. You may read chapter 17, regarding recreational aerobatics, for some ideas on how to amuse yourself with your new skills, and how to keep them current. Keep a good lookout, be safe, observe the FARs, and have a lot of fun!

If you are the competitive type and want to develop your aerobatic abilities to ever higher levels, turn to chapter 18.

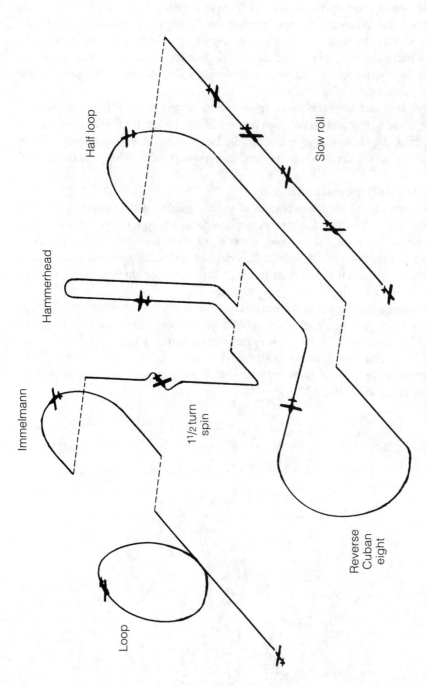

Half loop

Slow roll

Hammerhead

Immelmann

1 1/2 turn
spin

Reverse
Cuban
eight

Loop

Fig. 16-1. *A typical aerobatic sequence that is ideal for the pilot who has become competent in flying individual basic maneuvers.*

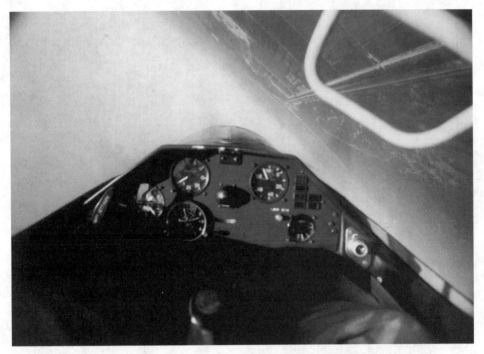

Fig. 16-2. *Be sure to have good ground reference points and lines.*

Fig. 16-3. *The G meter tells the story.*

Coffee, tea, or a roll?

The Boeing Airplane Company revolutionized air travel with the 707 intercontinental jetliner. The four-engine 180-passenger jet is still in service with quite a few of the smaller airlines throughout the world, and its KC-135 military tanker and AWACS versions still serve with the armed forces of the United States and several other nations. Until 1988, the American president's airplane, Air Force One, was a highly modified Boeing 707. Its distinctive lines are readily recognizable in its stubbier cousin, the 737, which is still one of Boeing's biggest sellers.

The 707's illustrious career was by no means assured when its prototype was unveiled back in 1954. Boeing was taking a big business gamble and the company's ex-

ecutives were going all out to promote the airplane. Bill Allen, Boeing's president, had invited a yacht-load of aviation industry officials to the 1955 Gold Cup Hydroplane Boat Race being held on Lake Washington, just outside Seattle, and had arranged for the 707 prototype to make several low passes over the event. It was the perfect public relations opportunity, the airplane's first appearance in front of a large crowd. More than 200,000 spectators, most of whom had never seen a passenger jet, watched in great excitement as the 707 rushed toward them, barely preceded by the ear piercing shriek of its jet engines. What they were about to see would exceed their wildest expectations and horrify Bill Allen.

With test pilot Tex Johnston at the controls, the 707 was streaking along at 450 mph at about 400 feet. Suddenly its nose pitched up sharply to 35° and to everyone's utter shock, it started rolling to the left. Shock turned to amazement and then to wild enthusiasm as the 707 majestically completed a giant, lazy aileron roll that would have done any Stearman pilot proud.

It swung around for another pass and delighted the crowd with another aileron roll right over Bill Allen's yacht. Allen was anything but delighted. He was too stunned even to be angry, and is remembered to have demanded a fistful of pills from one of his distinguished guests who was known to carry them for a heart condition.

When Tex Johnston was called on the carpet, he patiently explained that the aileron roll is the tamest of positive-G maneuvers. He had gently pulled a positive 1G and held it all the way around. There was never any danger of the fuel and hydraulic flows being interrupted due to negative G, and the airplane's low negative-limit load of -1G was irrelevant to the aileron roll. Nor was there a danger of any altitude loss at the end of the low-altitude maneuver because if anyone knew how to fly a proper aileron roll every time, he did.

"That airplane never knew it wasn't flying straight and level," Johnston said. Anyway, he would never do anything he didn't know he could do. No, it wasn't the first time—on an earlier occasion Mount Rainier was the only witness. He was almost fired, but in the end Allen let him off the hook. Things were more loosey-goosey then—if you wore snakeskin boots and a 10-gallon hat and were Boeing's best test pilot.

Johnston never understood what all the fuss was about. For him, the barrel roll was what a standard rate turn is to an instrument pilot. He was just trying to sell airplanes, he claimed, by showing what they could do. He wouldn't get off so lightly today, and reason tells us that's just as well. Or, is it?

17
Recreational aerobatics

THE VAST MAJORITY OF AEROBATIC PILOTS PRACTICE THE SPORT JUST FOR fun. Many of them don't belong to any formal support organization such as the International Aerobatic Club, and even in the IAC only about 10 percent of the members fly in aerobatic competition. Most recreational aerobatic pilots feel that in some ways they get a much greater sense of freedom from aerobatics than either competition pilots or airshow display pilots.

They experience the irresistible thrill that comes from being able to range at will through all three dimensions, but they are also free from the pressures and constraints of competition and airshow flying. They don't have to worry about aerobatic boxes, the fractional differences in control movements that make or break a high score, the opinions of judges, the tyranny of time slots, the substantial expenses of a sufficiently competitive or exotic airplane, and demanding schedules.

Above all, the recreational pilot is completely free to choose the time, the duration, and the maneuvers (within capabilities) for every flight. Recreational aerobatics is perhaps the closest we can come to a stereotypic image of aerobatics, the pleasant business of rolling and looping some colorful machine about the skies with complete abandon on a perfect day entirely for our own amusement.

Recreational aerobatics implies that your flying will be less intense because your maneuvers will be more basic than is required by competitive or airshow flying—if not, then you might as well compete or join the demo circuit. But to be safe and to keep the sense of fun alive, you have to maintain a set of recreational aerobatic standards. This task can be a real challenge because beyond the absolute minimum requirements imposed by the FARs, your standards are self-imposed. How best to keep current, what aircraft to choose, how to exercise quality control, how to expand your repertoire of maneuvers, and how to keep motivated, are among the many questions you will have to answer.

KEEPING CURRENT, STAYING SAFE

You have done your 10 hours of aerobatic dual in a Decathlon, maybe even a few hours in a two-seat Pitts, and you can hardly wait to get out there and enjoy yourself on your own. What you will soon notice is that you are, in fact, on your own. Nobody will tell you how much or how little to fly, what maneuvers to perform, and how to stay safe. It is *all* up to you. Here are a few ideas that might help you devise the standards to best suit your own circumstances.

Keeping current

A currency standard widely believed by aerobatic instructors to be a good minimum is two aerobatic sessions a month. It is advisable to fly your entire repertoire during each session. If you fly each basic maneuver covered in this book, the aerobatic portion of each flight should take you about 45 minutes. If you feel you could improve the way you fly a particular maneuver, make it your business to take the extra flight time to practice it, but not at the expense of the other maneuvers. If, for example, your Immelmann technique needs some polishing, make the time to practice it, but be sure to fly all the other maneuvers you planned, even if you have to fly an extra session or two. As is the case with all forms of flying, continuity is important to keep your skills sharp. Shorter, frequent flights are more productive than longer, infrequent flights.

It will not always be possible to maintain your schedule of two sessions a month, but you might still feel safe to fly aerobatics after a somewhat longer interval. In that case, you might want to be especially conservative in working up to the more complex maneuvers. You might find it useful to spend some extra time on simple rolls and loops before cautiously moving into combination maneuvers.

It is advisable to set yourself an absolute time limit beyond which you will not fly any aerobatics without a competency check with a qualified aerobatic instructor. Depending on your skills and experience, 45 or 60 days might be a reasonable absolute limit.

Recurrency checkride

As a recreational aerobatic pilot, your flying skills will be rarely evaluated by anyone, much less someone who knows anything about aerobatics. Over time, all sorts of bad habits might sneak into your aerobatic technique in spite of your best efforts. Regardless of how current you are, it is a good idea to get a reliable independent evaluation of your aerobatic skills from time to time. The best policy is to arrange a periodic recurrency checkride, much in the manner of a biennial flight review. Depending upon the amount of aerobatics you do, you might want to take a checkride with a qualified aerobatic instructor once every 12 or 24 months.

Staying safe

Beyond keeping current, the general safety standards of your recreational aerobatic flying is also largely in your own hands. It is your responsibility to observe all FARs and set yourself commonsense safety standards beyond legal requirements. Consider these points:

- Conduct every preflight as carefully and in as much detail as you were taught to do in training.

- Though not required by the FARs, wear a parachute even when flying solo.

- Be conservative in the altitudes you select for recreational aerobatics: fly high.

- Fly clearing turns scrupulously and frequently. Strive to see and be seen.

- Do your best to select practice areas where traffic density is low and times of day when traffic is generally light.

- Fly only those maneuvers that you were taught to do. If you want to learn new maneuvers, it is time for more dual instruction.

- From among the maneuvers in your repertoire, fly only those maneuvers that are specifically authorized in the aircraft's flight manual.

- Plan each maneuver and sequence prior to flying it and stick to your plan. Don't change a sequence or maneuver in midflight, except to skip or abort a maneuver in the interest of safety.

- Know your physical endurance limits. Know when to stop. Fly only when you are in good physical and mental condition. Remember, you are in it for fun.

- Make it your business to get a thorough aerobatic checkout in every new aircraft type you fly. If the new type is a single-seater, arrange the checkout in a two-seater of comparable performance and handling characteristics.

- Systematically keep informed of all airworthiness directives, service bulletins, and service difficulty reports for each aerobatic aircraft you fly. Don't rely on word of mouth. Subscribe to an information service or get permission to peruse someone else's.

- Keep abreast of changes in regulations and their applicability to aerobatic flight. Regulations might apply indirectly, in which case it is your responsibility to interpret them correctly. Seek informed guidance if at all in doubt.

Ultimately, safety is mostly a matter of common sense and a feeling of responsibility. Bear in mind that in the United States the authorities usually favor self-regulation (the avoidance of unsafe practices through responsible behavior) and regulate only when the system breaks down, when irresponsible behavior becomes sufficiently widespread to cause a problem for the aviation community and the community at large. It is our collective responsibility to fly safely and keep the pressures for the regulation of aerobatics to a minimum.

CHOICE OF AIRCRAFT

Most recreational aerobatic pilots have other aviation interests beyond aerobatics. More often than not, the handful of pleasant basic aerobatic maneuvers they want to fly from time to time is only one of their many flying activities. For such pilots it makes little sense to get a single-purpose, high-performance aerobatic machine. They will never learn to even scratch the surface of its capabilities, let alone fly it as it should be flown. It would be like buying an Indy 500 race car just to commute to the office, hit the grocery store, and occasionally run it up to the speed limit on a family outing.

The recreational aerobatic pilot is usually far better off getting an airplane that nicely performs basic and moderately advanced aerobatic maneuvers and is suitable for a whole range of other flying activities, many of them social. A popular requirement of recreational aerobatic pilots when selecting an airplane is more than one seat.

Your choice is widest if you don't plan to fly any negative-G maneuvers. In that case, you won't need an inverted system. You can spin, loop, aileron roll, and barrel roll your way across the land in such budget Sunday Loopers as the Cessna Aerobat, the Citabria (if you don't mind the excessive stick forces), and even the aerobatic version of the Bonanza. When you have your fill of loops and rolls, you and your passenger(s) can be off on a cross-country jaunt of a respectable distance, head for the beach, or do the airport coffee shop rounds.

If you want to do the full range of basic maneuvers you learned—including negative G maneuvers requiring an inverted system—and also want comfortable cross-country performance, the staple basic aerobatic trainer, the Bellanca Decathlon, or a low-wing Mudry CAP 10 are excellent choices.

A popular option for recreational aerobatic pilots is the world of antique biplanes. Practically all of them are capable of basic aerobatics (and some a lot more) and they are all living relics of aviation's most romantic period. If you want to relive the old barnstorming days, keep aviation history alive for all to experience, and fly aerobatics the old-fashioned way, nothing beats the antique biplane. Be prepared, though, for hefty maintenance bills and hangar fees to give the old birds the care they deserve. Stearmans, Wacos, Great Lakes, and Tiger Moths are among the choices to fly off into the silk scarf days.

If you want more muscle and zip, you have a variety of versatile options with a military flavor, among them the Beech T-34, the Italian Siai Marchetti SF .260, the Russian Yak 52, and the British Slingsby Firefly and its predecessors. These airplanes were designed for dual civilian and military trainer roles and are excellent at performing the aerobatic maneuvers taught in basic military flying courses. In the right hands, they can introduce the novice to an exciting special niche of aerobatic flight: combat maneuvering. They are also swift, fully IFR cross-country machines.

Perhaps the ultimate recreational aerobatic aircraft is the aerobatic warbird. The most popular entry level warbird is the North American T-6 Texan. (Do not think for a nanosecond that your 10 hours of aerobatics in a Decathlon or even a Pitts qualify you to do anything more in a T-6 than sit in it on the ground.) Warbirds capable of aerobatics are generally very high-performance aircraft designed for the ultimate winner-take-all competition. They are a heck of a lot of fun to fly, but pilots aspiring to fly them should get the thorough, lengthy, and conservative training they demand.

Sure, your basic Decathlon or Pitts aerobatics will be of some use when you begin to fly warbird aerobatics, but only to the extent that you will be familiar with some basic concepts regarding what you are trying to do. If you are unfamiliar with the warbird world and would like to know more, contact the Warbirds of America, EAA's warbird suborganization.

As of this writing you can get a thorough and safe introduction to flying the ultimate World War II fighter, the North American P-51 Mustang. The Stallion 51 Corp. of Kissimmee, Florida, and Nashua, New Hampshire, rents a fully dual-control P-51 Mustang in which you fly the airplane (including aerobatics) under the supervision of a qualified instructor. It will be the ride of your life. For more information, *see* Mustang Dogfight elsewhere in this book.

Mustang dreams aside, if you are a more ambitious recreational aerobatic pilot, you might find such high-performance aerobatic aircraft as the Pitts and the Christen Eagle irresistible, but to get your money's worth, be prepared to do a lot more in them than Sunday looping.

Whatever your choice of aircraft, be sure it has the ability to do all the aerobatic maneuvers you plan to fly, and get a thorough aerobatic checkout in type before you set out on your own.

PASSENGER CONSIDERATIONS

As a recreational aerobatic pilot, you will find no shortage of takers for an extra seat in your aircraft. Many of your passengers might have some aviation experience, many might not; few are likely to have any real experience in aerobatics. All who fly with you will put their full trust in you to safely fly them through an exciting aerobatic routine and return them to earth. It is a big responsibility.

The first cardinal rule of flying aerobatics with passengers is to never ever fly an aerobatic maneuver with them with which you are not 100 percent comfortable. You might find this point obvious, but you would be surprised by how many pilots are out there who cheerfully tell an apprehensive passenger—as the airspeed builds alarmingly—"Well, I haven't done this maneuver in five or six years, but really, it is just like riding a bicycle."

When you fly any passenger for any purpose you might be seen as the fearless barnstormer in the leather jacket and the white silk scarf or the highly trained professional with an impressive ration of wings, epaulets, and scrambled eggs on various articles of clothing, but in reality you are nothing more than their limo driver for the duration of their flight. You are providing them a service and you must do your best to give them exactly what they want: not what you think they should want, but what they really want. During every flight with a passenger you are aviation's ambassador. Your passengers' opinion of our aeronautical world will be greatly influenced by your conduct. It is easy to give us all a bad name. It is just as easy to make their flight one of the greatest experiences of their life.

Any passenger who has had little or no aerobatic experience is going to be apprehensive before an aerobatic flight. So, please, no macho garbage. Ever! The fewer surprises they have, the more secure they will feel. Two good techniques work well together to put passengers at ease prior to takeoff and in flight:

- Brief them in simple and clear detail on everything you are going to do maneuver by maneuver and what they are likely to feel as a passenger. Give them a good safety briefing, including use of the parachute without being alarmist about potential problems. Tell them in flight where to look in each maneuver to minimize the chance of queasiness. Straight ahead is best for most rolling maneuvers. Make them search with you for the horizon at the top of looping maneuvers and have them look at the wingtip in hammerheads.

- Explicitly tell them to instantly let you know if they are feeling the least bit unwell, at which point you will immediately fly straight and level. Show them where the airsickness bag is, tell them that you have an extra one, and brief them on operating the air vents.

Ask passengers after every maneuver how they feel, and tell them what's coming next. You will soon sense if you have to go easy or if your passenger is a natural.

Keep in mind that someone who hasn't flown aerobatics has a much lower tolerance for aerobatic maneuvers than you have. Remember how exhausted you used to get in the beginning from a few simple maneuvers. Tailor the ride to your passenger's stamina, not your own.

If your passenger feels ill, immediately transition to straight and level flight and open the air vents. If the passenger has any flying experience, let the passenger handle the controls. It is the most effective way to settle a queasy stomach. The rider might feel better if you talk about a pleasant topic that is unrelated to flying. A sip of water can work wonders, but if you do carry a plastic water bottle on an aerobatic flight, be sure the container is leakproof and the container's storage area is fail-safe. See if there is sufficient improvement in your passenger's condition to continue aerobatic maneuvers, but don't push it. Let them make the decision. Be aware that sometimes a passenger will manage to hold on until you have the field made, only to lose it as you are rolling down the runway.

The more considerate and accommodating you are to your passengers, the more rewarding they will find their flight with you, regardless of their previous flight experience.

KEEPING INFORMED

Recreational aerobatic pilots can find themselves so much on their own that it might seem difficult to keep up with developments in the world of aerobatics. Yet it is important to know what is going on. Safety issues arise, regulations change, and new attractive opportunities might emerge. If you know how to go about it, you need little effort to become part of a structured aerobatic support group even if you fly only recreational aerobatics.

The first step is to join the International Aerobatic Club, described in detail elsewhere in this book. Their excellent monthly magazine, *Sport Aerobatics*, is alone worth the annual dues and will keep you fully informed on aerobatic developments. A series of technical tips is also very useful.

For some recreational aerobatic pilots, membership in the IAC is sufficient, but you can find further support in your local IAC chapter (the IAC provides a list of chapters nationwide). At monthly chapter meetings, you can meet other aerobatic pilots in your area and become an active member of your local aerobatic community. The IAC and its local chapters can also help you with keeping track of all the airworthiness directives, service bulletins, and service difficulty reports applicable to your aircraft.

J.R. Campbell/Pompano Air Center

Fig. 17-1. *An antique open cockpit aerobatic biplane is an excellent choice for the recreational aerobatic pilot, but be sure to start with plenty of dual in type.*

Fig. 17-2. *A perfect afternoon.*

Fig. 17-3. *The Slingsby Firefly is a favored recreational aerobatic airplane in Europe.*

18
Competition aerobatics

UNLIMITED COMPETITION AEROBATICS IS WITHOUT A DOUBT THE HIGHEST form of aerobatic flying. Competitors are expected to be accomplished aerobatic pilots capable of flying all the maneuvers required by the standards of the competitions they enter. The objective is perfection, the flawless execution of the assigned compulsory or chosen freestyle maneuvers. Maneuvers are defined to exacting standards and given degrees of difficulty. The competing pilots' performance is judged by a scoring system that takes both of these factors into consideration.

Competition aerobatics is the best way to fine-tune and further develop your aerobatic abilities. There is no better aerobatic pilot than a top unlimited competitor, and getting started is relatively easy. Anyone who has completed the basic aerobatic course presented in this book and has taken some time to put together and practice a few basic aerobatic sequences is ready for entry-level competition. In most countries, aerobatic competition flying is organized to accommodate various levels of experience from basic to unlimited, and is usually overseen by some sort of a national aerobatic association.

Aspiring competition pilots start out by entering grass roots-level beginner's events and can work their way up to national unlimited competition flying. The next step is international competition and its ultimate event is the world championships held every two years. The pilots selected to represent their country in international competition are chosen from the ranks of the top national unlimited competitors.

Another segment of competitive aerobatics that was once the norm worldwide and is again becoming popular is the individual cup event such as the Breitling World Cup and the Fond du Lac Cup. Cup competition might be invitational and might feature corporate sponsorship.

This chapter introduces you to getting started in aerobatic competition. You learn how to fly your aerobatic sequences in the aerobatic box, how to select and enter your first event, what to expect, how to prepare, and how to choose the airplane you will fly. If you enjoy your first few events and want to make serious progress as a competition pilot, this chapter explains how to set up a regular training program and arrange for coaching.

We discuss what it takes to move on to higher levels of competition, based upon your results at this level of competition, your training regime, your time commitment, and your aircraft and financial considerations.

Before we get into the details, here is a brief look at how aerobatic competition began and how the competition scene progressed to the present.

ORIGINS

As soon as Pegoud and his contemporaries began to experiment back in 1913 with what an airplane could really do, aerobatics became a feature attraction of airshows and other aeronautical gatherings on both sides of the Atlantic. But aerobatic displays were mostly a titillating filler act for the main event in those early days, the quest for speed. The few stand-alone aerobatic displays that were organized were put on for their stunt value, attracting large, voyeuristic crowds that were interested in whether or not the crazy pilot was going to crash. The maneuvers flown at this stage were not well developed. Loops were the big favorite, popularized by the well-known American stunt pilot Lincoln Beechey and his Little Looper, and Katherine Stinson, the first woman to fly the loop, but the loop shape was often quite elliptical.

Fundamental maneuvers were perfected in the aerial arena of World War I. As barnstorming, airshows, and air racing blossomed in the postwar years, aerobatic display flying became a major attraction and increasingly sophisticated combination maneuvers began to appear. The best display pilots became sought-after national celebrities who never ceased in their attempts to outdo each other. Some form of formal competition was just a matter of time.

The Europeans were the first, in the early 1920s, to work out ways to evaluate and rank individual aerobatic performances. With their penchant for bureaucracy, they quickly concocted vast catalogs of maneuvers and intricate scoring systems, and turned aerobatic display flying into contests of compulsory and freestyle events throughout the continent.

The more individualistic Americans preferred to twirl and tumble across the skies in the old, free spirited way and let the airshow crowds be the judge through their pocketbooks. It was not until 1932 that America's first judged precision aerobatic competition was organized: the Freddie Lund Trophy. It was named after one of the country's premiere display pilots of the 1920s, who was popularly known as "The Man Without Nerves." It was an invitational contest for professional airshow display pilots and, characteristically, it consisted only of freestyle events.

In 1934, the first world championships were organized in Paris to great success. Various judged trophy events continued until World War II put sport aerobatics on hold for almost a decade. The Gulf Trophy led a revival in judged precision aerobatic competition in the United States in the immediate postwar years that lasted until 1952 when organized competition flying in America went into low-gear once again.

In 1962, under the auspices of the Aerobatic Club of America and organized by the legendary Duane Cole, the first officially sponsored U.S. National Championships were held in Phoenix, Arizona. It was from this new beginning that the present structure of local and national aerobatic competitions developed in the United States under the supervision of the International Aerobatic Club (IAC), which largely assumed the role of the Aerobatic Club of America in 1970 and is a division of the Experimental Aircraft Association.

European aerobatic competitions were quickly reestablished in postwar Europe, with the exception of Germany, where flying was banned for several years. The era's premiere international event became the annual Lockheed Trophy competition in Britain; the first contest was in 1955.

The Lockheed Trophy of 1957 changed the world of aerobatic competition forever, with the appearance of the Czech Zlin 226 Trener Master. The Czech team's agile monoplanes introduced a whole new performance style that was heavy on outside and autorotational maneuvers, which came to be known as *dynamic aerobatics*, and is the standard to this day.

The Czechoslovaks, flush with success, petitioned the Federation Aeronautique International (FAI) in 1959 to be allowed to hold the first world championships officially sanctioned by the international organization. The request was granted for 1960, and the world championships have been held every two years ever since.

The next important development was the acceptance of the *Aresti* aerobatic shorthand and scoring system for the world championships in 1964. The system standardized the definition and evaluation of aerobatic maneuvers in international competition and filtered down in some form or fashion into national and local use throughout the aerobatic community.

American pilots had been at the world championships right from the beginning in 1960 when Frank Price and his Great Lakes represented the United States on a shoestring budget out of his own pocket at great personal sacrifice. The next championships in Hungary saw the participation of three Americans, Lindsey Parsons, Duane Cole, and Rod Jocelyn. Parsons placed a very respectable fifth in his Great Lakes biplane, ahead of 18 Zlins and four Yak 18s.

American competitors persevered without the benefits of the massive government financial, organizational, and training support that was enjoyed by the eastern Europeans and the Russians. Bob Herendeen caused a sensation in 1966 at Moscow with the first appearance of the tiny overpowered Pitts S-1C.

Ten years after the modern world championships began, Herendeen placed second in a Pitts S-1S and Charlie Hillard was third in his Spinks Akromaster. Other U.S. team members placed well enough in that 1970 contest to capture the world championship team title for the United States. America's Mary Gaffaney placed third in the women's competition, flying a Pitts S-1S. This period also saw the end of the Zlin Trener era because the design was increasingly outclassed by the Pitts and a growing variety of monoplanes with better performance.

Americans captured the men's and women's titles in 1972 in France; Charlie Hillard and Mary Gaffaney flew Pitts S-1S aircraft. American pilots have since consistently been among the top world competitors, winning a respectable share of top honors and eventually exchanging the loyal little Pitts for advanced monoplanes.

Under the IAC's management and with generous grass roots financial support from private donations to the United States Aerobatic Foundation, the continuing participation of America's top pilots in world competition is assured.

INTERNATIONAL AEROBATIC CLUB

Aerobatic competition in the United States is organized under the auspices of the International Aerobatic Club, a division of the Experimental Aircraft Association. Two types of competition are regional events and the annual National Aerobatic Championships. Regional events are sanctioned by the IAC's national office but a local IAC chapter is entirely responsible for each event. The national office organizes and runs the annual national championships.

Men and women compete together in America—there are no separate women's championships as in many other countries and at the world championships. Patty Wagstaff made aviation history in 1991 by becoming the first woman to win America's top aerobatic event, the Unlimited U.S. National Championship. She took the title again in 1992 and 1993.

Competitions are held in five categories: *basic*, *sportsman*, *intermediate*, *advanced*, and *unlimited*. The gradually increasing level of complexity from basic to unlimited allows pilots of all experience levels to compete safely and provides the novice with a path to the highest levels of competition flying. All categories except basic are held at national competitions. Regional competitions in a specific category require sufficient entrants. More and more regional competitions are able to regularly field all five categories due to the increasing popularity of competition flying in recent years.

All categories have *known sequences*, which means that the pilots know the sequence well in advance. The intermediate, advanced, and unlimited categories also have an *unknown sequence* and a *freestyle sequence*. The unknown sequence is presented to pilots at the competition only 12 hours before the event; practice is prohibited. The freestyle sequence is composed by each pilot. (A freestyle sequence is an option in the sportsman category.)

The IAC plays an important role in putting together the known and unknown sequences for each appropriate category. One common known sequence per category for each year is made public at the beginning of the year and is flown at all contests nationwide throughout the competition season. The unlimited known sequence is developed in collaboration with national aerobatic authorities around the world under FAI auspices and is the same for a given year worldwide.

At the moment pilots may enter any category at both the regional and national level. There are no prequalification requirements, though at the national level some selection process might become necessary in the future if the number of entrants increases to unmanageable levels. It is up to the pilot's judgment to decide what level is most appropriate based upon flying abilities; the system has worked well. Pushing too hard too soon could be counterproductive because it might mean getting into complex maneuvers without having had enough experience with the maneuvers that are the foundation of good performance at all levels.

A closer look at each category

Basic. This category has been recently introduced and its sequence consists only of a spin, loop, and roll in that order. The sequence is flown and scored twice in each competition. It provides an opportunity for the total novice to participate in competition flying without any real pressure or need for expensive equipment.

Sportsman. This category is the most popular entry-level competition event. Though it is quite a bit more demanding than the basic category, it can be flown by anyone who has learned the maneuvers presented in chapters 4–14 and is the foundation event for competitive aerobatics. (A subsection of this chapter discusses how to prepare for and fly your first competition, starting with the sportsman category.)

Sportsman contests consist of two sequences at regional competitions and two or three sequences at the nationals. The first flight (as well as the second at the nationals, if there are three sequences) is the known sequence. For the second flight (third at nationals), the pilot may fly the known sequence again or may fly a freestyle sequence. Most pilots opt to repeat the known sequence.

Intermediate. More complex forms of positive maneuvers are introduced at the intermediate level. Pilots encounter the unknown sequence for the first time and are also required to fly a freestyle sequence. Increasing emphasis is placed upon vertical maneuvers; hence, to do well it is important to fly an airplane with good vertical performance.

Advanced. Pilots encounter negative maneuvers for the first time in this category. Also more complex variations of intermediate maneuvers and multiple combination maneuvers. Sequences are known, unknown, and freestyle.

Unlimited. The Unlimited category is the Indy 500 of aerobatics. All the stops are out. Only the most skilled pilots flying the highest performance aircraft need apply. New maneuvers are tailslides, outside snap rolls, and vertical upline snap rolls. The pace is punishing. Good physical condition is essential.

Unlimited sequences are known, unknown, and freestyle. In addition, a separate fourth event, a 4-minute freestyle, is judged independently under its own set of rules and is not part of the competitors' final total score. The 4-minute freestyle is really a display performance, similar to an airshow sequence, that was primarily established to allow pilots to fly maneuvers that are hard to judge and not otherwise flown in competition: Lomcovak, torque rolls, and the Bessenyei.

If you are thinking of flying your first competition and can fly all the maneuvers, you have to learn one more essential skill before you sign up: how to read the sequence cards.

AEROBATIC SHORTHAND

Catalogs of aerobatic maneuvers were developed as early as the first competitions of the 1920s, when it was common to fly sequences with as many as 20 maneuvers. Because the maneuvers were becoming so complex, it became more and more difficult to remember sequences and there was little room in the cramped cockpits for voluminous pilot's notes and even less time to read them in midperformance. Judges also found it increasingly difficult to track and judge fast-paced sequences. Clearly, there was a need for some form of shorthand, a written language and scoring system of aerobatics that could be easily displayed in the cockpit as well as the judges' positions, and tracked at a glance in the air and on the ground.

Following the early efforts of France's D'Huc Dressler, the solution was provided by José Luis Aresti of Spain, whose name has become synonymous with the aerobatic shorthand and scoring system that he developed. Aresti, a test pilot and flamboyant aerobatic performer who was a virtuoso with the Bücker Jungmeister, took it upon himself to devise a catalogue of aerobatic maneuvers, the *Sistema Aresti*, which came into use in Spain in 1961, one year after the formal FAI-recognized world championships began. When the world championships were held in Spain in 1964, the *Aresti Catalogue* was accepted by the FAI as the standard. It has been with us ever since, officially known today as the *FAI Aerobatic Catalog*, updated to include newly developed maneuvers and modified to reflect the effects of improvements in aircraft performance on the difficulty of flying certain maneuvers (FIG. 18-4).

Two aspects of the system need to be understood: how the maneuvers are depicted and how they are valued. The fundamental technique of depiction is easy to understand in concept and needs some study to grasp the details. Every maneuver is based on a line, representing the line of flight. A dot at the beginning of the line represents the beginning of the maneuver, a short perpendicular line at the end signals the completion of the maneuver. For some figures, such as the loop, the line of flight alone is sufficient to depict the maneuver. Other maneuvers are identified by symbols and numbers placed on the line of flight.

A solid line of flight signals positive acceleration loading (Gs) on the aircraft. A dotted line indicates negative loading on the aircraft; thus, upright straight and level is a solid line, inverted straight and level is dotted. A vertical pull-up followed by a pushover into a vertical downline is represented by a solid line that goes up, a dotted line for the pushover (even though the aircraft is upright relative to the ground, it is experiencing negative G), followed by a solid line that goes down.

Some maneuvers stand alone, others are built by the user of the catalog. For example, a simple inside loop might stand alone, depicted by the loop figure taken directly from the catalog. If a roll is inserted at some point on the loop, the roll symbol (from the catalog) is placed at the appropriate spot on the loop figure.

Study the Aresti symbols again for a basic understanding of how the maneuvers are depicted and then wade into the catalog with your instructor.

The catalog breaks down aerobatic maneuver elements into nine families, each containing an exhaustive set of variations of the element represented by the family.

The nine families are:

1. Lines and angles

2. Turns and rolling turns

3. Combination of lines

4. Spins

5. Stall turns (hammerheads)

6. Tailslides

7. Loops and eights

8. Combination of lines, angles, and loops

9. Rolls

Most figures within 1–8 may be flown stand alone or combined with a figure from 9 (rolls)—according to rules that are easy to understand—for greater complexity and a higher value. Figures in family 9 must always be combined with elements of another family. More than 15,000 distinct figures can be composed in this fashion.

Valuation is relatively straightforward. Each element has a number next to it that is a difficulty coefficient, a *K factor*. If the element is used alone, this coefficient is the value for the maneuver. If rolls are added to the element, then the K factor of each roll is added to the element's value for a total value. It doesn't take long to figure out that vertical maneuvers are generally valued the highest, so to maximize value, the trick is to choose vertical figures and load them down with difficult rolls. The total value of each maneuver is added together for a grand total value of the sequence.

The sequence is efficiently depicted on a card that is clamped to the instrument panel for easy in-flight reference. The card even indicates the direction of the wind and the position of each maneuver in the aerobatic box—the box is explained in the next subsection. See how easily you can decipher the Sportsman sequence presented in this chapter (FIG. 18-5).

YOUR FIRST COMPETITION

When the decision is made to enter your first competition, you have a variety of tasks to accomplish in preparation. You have to decide what aircraft to fly, you have to learn to fly in the aerobatic box, you have to become proficient at the known Sportsman sequence (we make the assumption that your first venture into competitive aerobatics will be in the Sportsman category), you have to learn the rules of competition, and it helps to learn to decipher the Aresti system.

Aircraft selection

There are no rules or restrictions on the aircraft you can fly in Sportsman competition, but from a practical standpoint it is best to avoid high-performance aircraft with capabilities that far exceed the demands of the category's fairly simple maneuvers. You are probably best off in a Decathlon if you plan to go no further than the Sportsman category or one of the lower powered Pitts or a CAP 10 if you are using Sportsman as an apprenticeship for higher level competition.

The aerobatic box and learning to fly in it

The aerobatic box is the competition arena (FIG. 18-6). It is a standard cube of airspace in which competition aerobatic sequences must be flown. It is 3,300 feet (1,000 meters) wide, 3,300 feet long, and its top is at 3,500 feet agl. The floor of the box is at different altitudes for different categories for safety reasons: 1,500-foot floor for basic and sportsman, 1,200-foot floor for intermediate, 800-foot floor for advanced, and 328-foot floor (100 meters) for unlimited. Length and width of the box are marked by white ground markers and positioning judges monitor the competitors' compliance with the limits. The sequence receives a scoring penalty if an aircraft exceeds the limits of the box.

The good news for the aspiring sportsman pilot is that the floor of the aerobatic box is the FAA's altitude limit for aerobatic flight, so selecting a practice box does not present a big problem and does not require any special waivers.

With your instructor's assistance, select a practice box by identifying two landmarks approximately 3,300 feet apart and using it as the main axis of the box along which you will fly your sequences. Road segments are easily measured, as are power lines. One of the best options is to find a 3,300-foot runway at a little used uncontrolled airport, position yourself off to one side to avoid traffic and use it as a reference.

Moving your maneuvers down to the lower altitudes required by the box takes some getting used to, but is fairly comfortably accomplished if done in stages. Fly each maneuver so that you would finish it at 2,500 feet instead of the 3,000–3,500 feet you are accustomed to in training. When you are comfortable, move each maneuver down another 500 feet. This is sufficiently low because you will find you can fly the entire sportsman sequence with a floor of 2,000 feet. The cushion of 500 feet enables you to stay in the box even if you inadvertently lose extra altitude in a maneuver. As you feel more comfortable with individual maneuvers in the box, start stringing them together, and you are on your way.

Training

When you have obtained and fully understand the Sportsman known sequence it is time to learn to fly it. Master the sequence in stages. Learn to fly the first three maneuvers well, then learn the next three, and then the rest. Fly the entire sequence only after you are proficient in its stages.

You need to practice two to four times a week to learn the sequence to competitive standards. The full sequence takes about 5 minutes. It is best to fly it twice per practice session and then practice the maneuvers you think need work. Plan for at most 30–35 minutes per session. Given the rigors of aerobatics, any more time than that is unlikely to be productive.

Practice the sequence in low wind, high wind, smooth days, choppy days, good visibility, not so good visibility, and at any time of day. You never know what conditions you will encounter in competition, so be prepared. A personal visibility limit in practice of 5 miles is a good idea for safety, though competitions are held right down to the 3-mile VFR visibility minimum.

When you become reasonably adept at flying through the whole sequence, it is time to fine-tune it. Arrange for an IAC judge or a highly experienced competition pilot to observe and critique your performance from the judges' standpoint, and work with them on correcting any deficiencies.

At the competition

Regional aerobatic competitions are usually held over weekends with Friday being a practice day. It is important to have the right attitude to get the most out of a competition. Yes, you do want to do well, to be competitive, but never lose sight of the fact that you are also there to have fun, to enjoy yourself, and to learn. Never be intimidated. Your fellow competitors started just like you and are also there to have a good time as much as to compete. Never be discouraged or embarrassed if you don't do as well as you would have liked. Learn from the experience. You will find you are welcome with open arms by your fellow competitors. The camaraderie of regional competitions is one of their most pleasant characteristics.

The first order of business when you arrive at the competition is to register. Your airplane will undergo a technical inspection by a contest official to assure the organizers that it is up to the task. Get on the practice list. Introduce yourself to some of the more experienced contestants if you don't already know them and ask them to critique your flight. You will invariably find them helpful and cooperative.

Your practice session is a good opportunity to check out the lay of the land. Note the position of the box and prominent landmarks.

The morning of the first event starts with a pilot briefing to explain procedures for the day and conduct a weather briefing. The pilots' contest positions are drawn from a hat and then it is time to go and fly. Relax, be safe, and concentrate, but enjoy yourself.

Ask to look at the judges' score sheets after your flight for a detailed analysis of how you did, beyond the final score. Pay particular attention to the "remarks" column. You might be able to photocopy the sheet for future reference. Be sure to take advantage of all the talent gathered at a contest. You can learn a lot by watching others. Don't be shy to ask for tips if you feel that some aspects of your technique need improvement.

Your first competition is a great milestone. It is the culmination of all your hard work, and it is also a beginning, the first step into an exciting new group of the aviation community. Whether you plan to remain in the Sportsman level or work your way up into higher categories, competition flying will without a doubt make you a better aerobatic pilot. And don't be surprised if you find it addictive.

BEYOND SPORTSMAN

It is hard work to become sufficiently competent to compete and do well in the Sportsman category, but the pace is a leisurely ramble through the skies compared to the demands of doing well in the categories beyond. Developing a level of proficiency at the Intermediate and Advanced levels leaves little time for other interests, and flying at the National Unlimited level is practically akin to being a pro athlete (alas, without the financial rewards of popular pro sports). The true reward of competing beyond Sportsman is the exhilaration of being among the best; of flying at a level that fewer and fewer pilots can match as you move up in the ranks.

How and when you decide to move beyond Sportsman can mean the difference between a frustrating experience or the most thrilling time of your life. The key is not making the move too early, not getting in over your head. Each category is an apprenticeship for the next one and you are best off conscientiously paying your dues. If you move up before there is nothing more for you to learn in the category you are competing, you will most likely lack the foundations required to do well in the next category. It is not uncommon to see over-enthusiastic pilots who hurried beyond Sportsman struggle with basic elements in advanced competitions that should be second nature to them.

Different pilots have different objectives, of course, and you should tailor your progress to your own particular goals. As a rule of thumb, don't move up before you get the sense that you are no longer learning in your present category; if you want to be competitive at the regional or national level, don't move up before you can consistently place in the top three in your present category at the regional or national level.

When you decide to move up, be sure to get dual instruction in the maneuvers new to you and be prepared to practice, practice, and practice, no excuses. Practice sessions three times a week are the norm for many serious competition pilots at the advanced levels, increasing to daily practice several weeks before a contest. You should also organize some form of periodic formal coaching. Diligently practicing your mistakes without the benefit of expert critique makes it difficult to improve. Many pilots band together from time to time for a week or two of joint practice under the guidance of an experienced coach.

Your airplane will also make more and more of a difference as you advance in the ranks. It is difficult at the more advanced levels to place in any aircraft type but the best, and the best is expensive. But the flying . . . wow!

In the end, to get the most out of your life and aerobatic competition, a well planned and steady progression up the ranks to the ultimate level of your abilities is the only way to go. And who knows? Place among the top five at the U.S. Unlimited Nationals the year before a world championship contest, and you will be on your way to the worlds.

Patty Wagstaff

Fig. 18-1. *The original Laser 200 has had its rewards.*

Mike Goulian Airshows/A. Parsons

Fig. 18-2. *In the box: A national champion hard at work.*

Fig. 18-3. *Awaiting your turn. Laser 200 at the U.S. Nationals, ready to go.*

Upright straightline flight
(positive G)

Inverted straightline flight
(negative G)

Slow roll

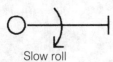

Half-roll to inverted

Outside loop
(from inverted)

Half loop to inverted

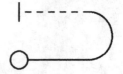

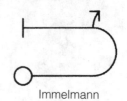

Immelmann

Fig. 18-4. *A sample selection of Aresti figures. The circle is the beginning of each maneuver and the short vertical line is the end.* Ron Burns

Fig. 18-4. *Continued*

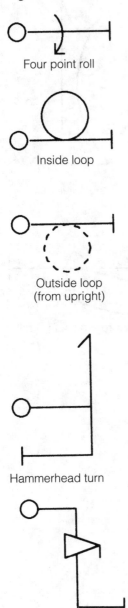

Four point roll

Inside loop

Outside loop
(from upright)

Hammerhead turn

One-turn spin

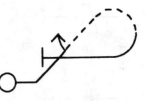

Half-Cuban eight

Reverse half-Cuban eight

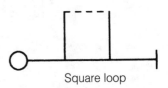

Square loop

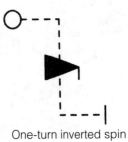

One-turn inverted spin

Fig. 18-4. *Continued*

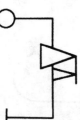

One-and-a-half
turn spin

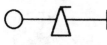

Inside snap roll

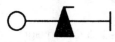

Outside snap roll

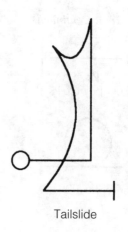

Tailslide

360° rolling circle; 1 roll

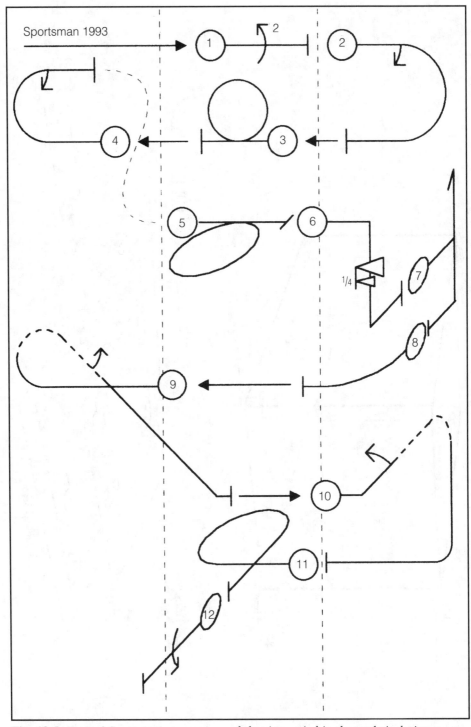

Fig. 18-5. *Typical Sportsman sequence card that is carried in the cockpit during a competitive flight.* Ron Burns

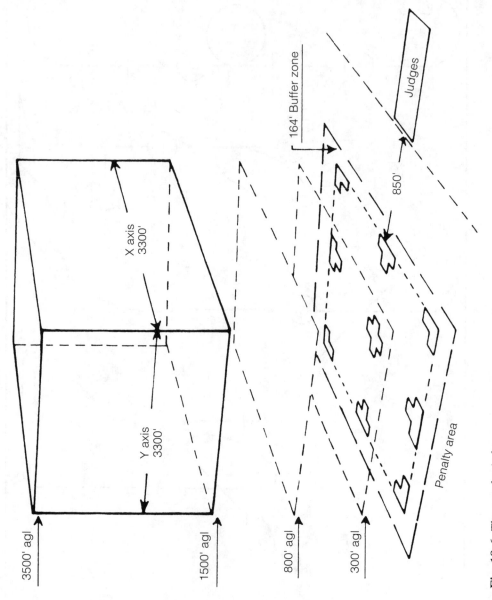

X axis
3300'

Y axis
3300'

3500' agl

1500' agl

800' agl

300' agl

164' Buffer zone

850'

Judges

Penalty area

Fig. 18-6. *The aerobatic box.*

Fig. 18-7. *A Decathlon is just as suitable for Sportsman competition as a Pitts Special.*

Fig. 18-8. *Sukhoi 29 cockpit.*

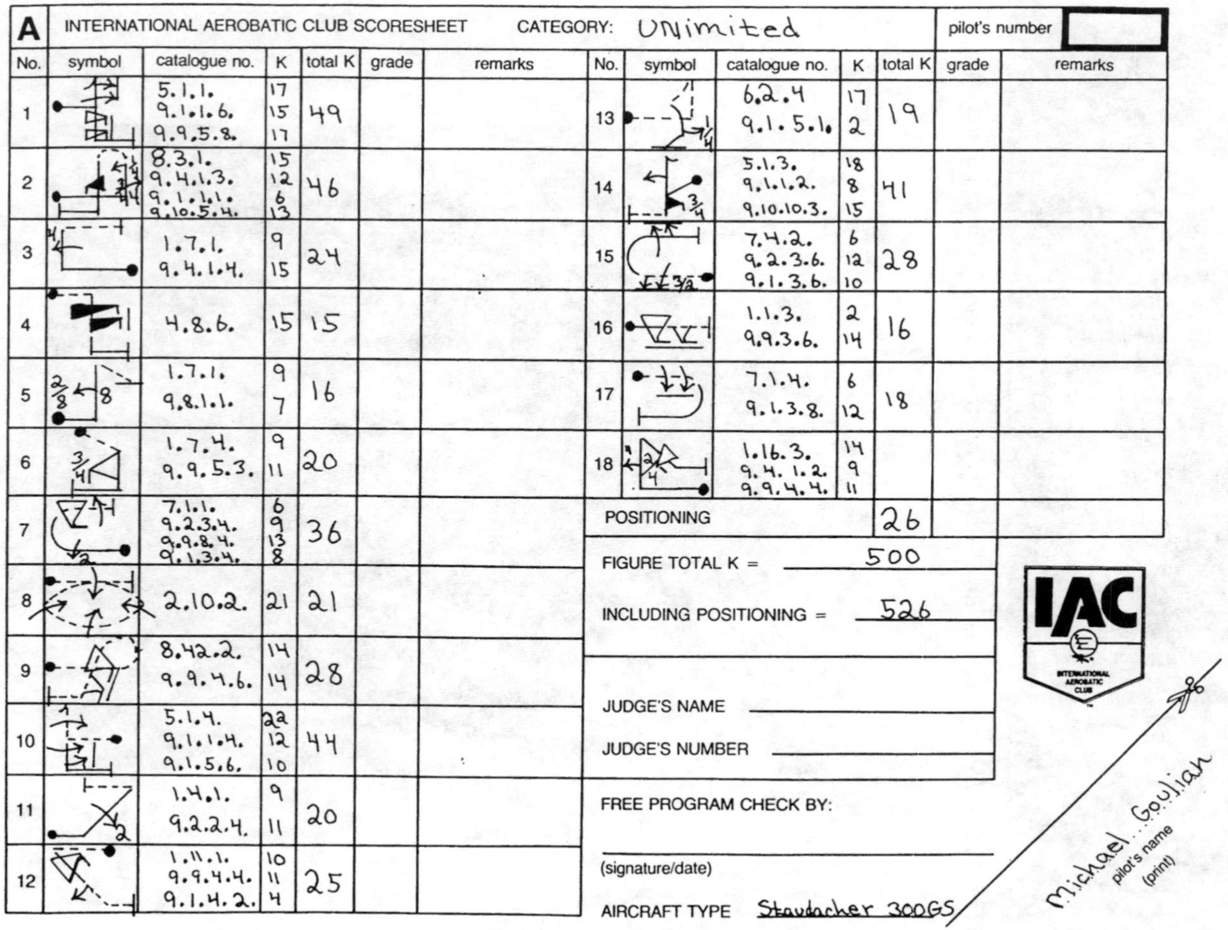

A INTERNATIONAL AEROBATIC CLUB SCORESHEET CATEGORY: Unlimited pilot's number

No.	symbol	catalogue no.	K	total K	grade	remarks
1		5.1.1. 9.1.1.6. 9.9.5.8.	17 15 17	49		
2		8.3.1. 9.4.1.3. 9.1.1.1. 9.10.5.4.	15 12 6 13	46		
3		1.7.1. 9.4.1.4.	9 15	24		
4		4.8.6.	15	15		
5		1.7.1. 9.8.1.1.	9 7	16		
6		1.7.4. 9.9.5.3.	9 11	20		
7		7.1.1. 9.2.3.4. 9.9.8.4. 9.1.3.4.	6 9 13 8	36		
8		2.10.2.	21	21		
9		8.42.2. 9.9.4.6.	14 14	28		
10		5.1.4. 9.1.1.4. 9.1.5.6.	22 12 10	44		
11		1.4.1. 9.2.2.4.	9 11	20		
12		1.11.1. 9.9.4.4. 9.1.4.2.	10 11 4	25		
13		6.2.4 9.1.5.1.	17 2	19		
14		5.1.3. 9.1.1.2. 9.10.10.3.	18 8 15	41		
15		7.4.2. 9.2.3.6. 9.1.3.6.	6 12 10	28		
16		1.1.3. 9.9.3.6.	2 14	16		
17		7.1.4. 9.1.3.8.	6 12	18		
18		1.16.3. 9.4.1.2. 9.9.4.4.	14 9 11			
POSITIONING				26		

FIGURE TOTAL K = 500

INCLUDING POSITIONING = 526

JUDGE'S NAME

JUDGE'S NUMBER

FREE PROGRAM CHECK BY:

(signature/date)

AIRCRAFT TYPE Staudacher 300GS

Michael Goulian
pilot's name
(print)

Fig. 18-9. *Judge's score sheet.*

Formation aerobatics

The lyrical aerial ballet of The French Connection aerobatic display team and the performances of such aerobatic formation teams as the Thunderbirds, Blue Angels, Eagles, and the Six of Diamonds are among the most thrilling airshow acts. They are also a highly specialized and difficult form of aerobatics. The formation aerobatic pilot must first become very good at formation flying and aerobatics separately. Only then can the two skills begin to be carefully meshed together.

The fundamental concept of formation aerobatics is akin to basic formation flying. One aircraft is the lead and all others form up relative to it. The lead pilot is the formation's maestro, responsible for flying the program and for cluing in the other formation members to what comes next throughout the flight. The other pilots' job is to perfectly stay in position, making the entire formation fly as if it were one single aircraft.

The lead pilot maneuvers relative to the ground. Every other formation member maneuvers strictly relative to the aircraft on which it is formed, which is the aircraft closest to it between itself and the lead. For example, in a symmetrical three-ship formation, each wingman *flies off* the lead. An aircraft joining this formation by forming on a wingman flies off that particular wingman, and so on.

The margin for error is practically nonexistent in a tight aerobatic formation because the wings are only 3 feet apart. At the high speeds of the aircraft, the slightest divergence in the wrong direction often leaves literally 1 second or less to react before a midair. It is therefore a cardinal rule of all formation flying that *all nonlead pilots look at all times at the aircraft on which they are formed*. All pilots must also be acutely aware of the position of the *parts* of their aircraft in the formation. Wings can too quickly get in the way when cockpits are still far apart.

The lead and nonlead pilots face equally difficult but quite different challenges. The nonlead pilots must anticipate every move of the aircraft on which they are formed and jockey power and controls to remain motionless relative to it. Take a simple turn to the left in a symmetrical three-ship formation: when the lead pilot turns, the inside wingman must also turn, slow down (the lead's inboard wing slows), and descend slightly to remain motionless relative to the lead (the lead's inboard wing dips down; therefore, the left wingman's outboard wing must be in a lower position to remain properly aligned). The outside wingman, on the other hand, has to turn, speed up, and climb slightly to remain motionless relative to the lead. Now imagine what it takes to safely fly a formation slow roll. . . . You can be sure that the more effortless a formation aerobatic act appears, the harder the pilots are hustling to earn their airshow fee.

The lead position has its own set of challenges. Experienced formation pilots often say that a formation is only as good as the lead is. A good lead pilot who intimately understands the effect of every move on the nonlead pilots, and makes proper flying adjustments to ease the burden, can make formation flying a pleasure. An inexperienced lead, whose airspeed, altitude, and timing is all over the skies, not only causes a high workload and a lot of grief for the rest of the formation, but can also make the flight unsafe.

For anyone planning to learn formation aerobatics, it is imperative to train with an instructor who has excellent credentials in aerobatics and formation flying, as well as formation aerobatics.

19
Airshow display aerobatics

AEROBATIC DISPLAY FLYING ON THE AIRSHOW CIRCUIT IS GREAT FUN, AND for the better display pilots it is a way to defray the costs of owning and flying their expensive aircraft. A handful of professional airshow pilots even manage to make a living from display flying. Though superficially similar to other forms of aerobatics, airshow display aerobatics is a highly specialized form of flying with its own set of specific skill requirements. In some respects, it is more difficult than competition flying, in other respects it is easier.

The greatest challenge for the airshow display pilot is flying safely in close proximity to the ground. To deliver the thrills the crowds expect, airshow aerobatics is flown at lower altitudes than any other form of aerobatics. The visual environment and its clues are different from the aerobatic environment at altitude and there is very little room for error or miscalculation. Aerobatic pilots of all skill levels, including top competition pilots, must be specifically trained for low-altitude aerobatics, regardless of other experience.

While performing at low altitudes is demanding, the technical level required of the individual maneuvers in airshow aerobatics is lower than in top-level competition flying. Airshow acts are designed to entertain. The crowd is the only judge, and in a technical sense not a very demanding one. To please the spectators, maneuvers do not have to be as complex or as perfectly executed as they must be for a high score in aerobatic competition.

THE AIRSHOW SCENE

Hundreds of annual airshows in the United States range from the grass roots open-house days at the local field to the mass migration to Oshkosh every summer. The supervising entity of all airshows in the country is the Federal Aviation Administration.

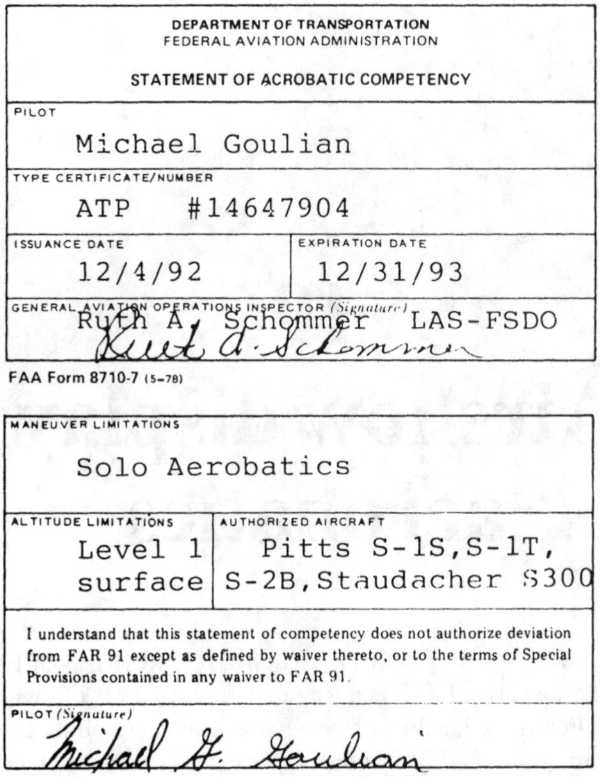

```
┌─────────────────────────────────────────────────┐
│           DEPARTMENT OF TRANSPORTATION            │
│          FEDERAL AVIATION ADMINISTRATION          │
│                                                   │
│          STATEMENT OF ACROBATIC COMPETENCY        │
├───────────────────────────────────────────────────┤
│ PILOT                                             │
│     Michael Goulian                               │
├───────────────────────────────────────────────────┤
│ TYPE CERTIFICATE/NUMBER                           │
│     ATP      #14647904                            │
├─────────────────────────┬─────────────────────────┤
│ ISSUANCE DATE           │ EXPIRATION DATE         │
│     12/4/92             │   12/31/93              │
├─────────────────────────┴─────────────────────────┤
│ GENERAL AVIATION OPERATIONS INSPECTOR (Signature) │
│   Ruth A. Schommer          LAS-FSDO              │
└───────────────────────────────────────────────────┘
FAA Form 8710-7 (5—78)

┌───────────────────────────────────────────────────┐
│ MANEUVER LIMITATIONS                              │
│     Solo Aerobatics                               │
├─────────────────────────┬─────────────────────────┤
│ ALTITUDE LIMITATIONS    │ AUTHORIZED AIRCRAFT     │
│     Level 1             │  Pitts S-1S,S-1T,       │
│     surface             │ S-2B,Staudacher S300    │
├─────────────────────────┴─────────────────────────┤
│  I understand that this statement of competency   │
│  does not authorize deviation from FAR 91 except  │
│  as defined by waiver thereto, or to the terms of │
│  Special Provisions contained in any waiver to    │
│  FAR 91.                                          │
├───────────────────────────────────────────────────┤
│ PILOT (Signature)                                 │
│   Michael F. Goulian                              │
└───────────────────────────────────────────────────┘
```

Fig. 19-1. *Pilot's waiver to perform at airshows.*

Any airport or organization wishing to hold an airshow must obtain an airshow waiver from the FAA for each individual event every time it is held. If aerobatics are to be performed, an additional low-level aerobatic airshow waiver must be obtained. Under the terms of an airshow's aerobatic waiver, pilots who hold FAA-issued low-level aerobatic waivers may perform aerobatics if they are named in the airshow's proposal to the FAA.

The coordinating organization for airshows nationwide is the International Council on Airshows (ICAS). ICAS does not sponsor or endorse any airshows. Rather, it is a clearinghouse of information for airshow organizers, performers and exhibitors who always contract with each other directly. The annual ICAS convention is an important forum in which each season's major airshows are announced, and trends and developments in the airshow business are discussed. Novice airshow organizers might obtain valuable advice and contacts from ICAS. ICAS also manages under FAA authority an *Aerobatic Competency Evaluator* (ACE) program to ensure some form of quality control over pilots authorized to fly airshow aerobatics.

All the big civilian airshows are basically big business. Even the so-called non-profit shows are run on a profit basis to benefit nonprofit causes or organizations instead of shareholders or investors. The basic idea is to take in more money from spectators and advertisers than it costs to put on the show.

Mira Slovak Aviation

Fig. 19-2. *The irrepressible Mira Slovak in his Bücker.*

The top airshow acts are paid a fee plus expenses. The fees depend upon the quality of the act and the experience of the pilot as well as the size and financial resources of the airshow. Less experienced performers generally have to start participating in the local grass roots airshows and should initially expect to be offered only their expenses as payment.

PILOT QUALIFICATION REQUIREMENTS

Let's suppose you just exchanged your straight-and-level four-banger for a Pitts Special. Your Uncle Harry has been watching you grind your way through basic aerobatics in the Pitts out of his grass strip and sees a great money-making opportunity. "Let's call your act the Little Ground Looper and charge the gum-chewing public $5 a head to come and feast their eyes," he says. You are flattered, but can you do it? No you can not. The splintered remains of aerobatic airplanes flown by those who beat your Uncle Harry to his idea have seen to it that airshow aerobatic pilots must meet certain FAA mandated standards of competence and rightly so.

Flying aerobatics at an airshow requires a low-level aerobatic waiver issued by the FAA that authorizes you to perform aerobatics below 1,500 feet. Depending on your skill and objectives, the waiver might be authorized for specific maneuvers only, or it might be unrestricted. Waivers have to be renewed every year.

To obtain a waiver, you must take an oral examination and perform aerobatics in front of an FAA aerobatic examiner or an ICAS-appointed aerobatic competency evaluator who then submits the paperwork to the FAA. A competency evaluator has to be an aerobatic instructor and a competition pilot or airshow performer of some repute. The FAA authorized ICAS to establish the ACE program to place the responsibility for evaluating aerobatic competence in the hands of those best qualified to do so.

A first-time waiver authorizes the holder to perform aerobatics down to 1,000 feet. At the examiner's discretion, in view of the applicant's skill level, the first waiver might be issued down to 500 feet. After the first year, the renewal waiver is issued down to 500 feet. After the second year, it is issued down to 200 feet and only after the third year is it issued down to the surface.

The ICAS ACE program allows a waiver to be renewed without having to fly in front of an examiner again or taking another oral examination if one of two conditions apply. First, you have performed aerobatics at eight airshows during the year or, second, you have flown in four airshows and had four of your aerobatic practice sessions during the year witnessed by an ACE specifically for the purpose of maintaining your waiver.

A waiver that is renewed by the FAA requires an annual flight and oral examination.

AIRSHOW DISPLAY FLYING SKILLS

Anyone aspiring to perform aerobatics in an airshow must be 100 percent comfortable with flying all the planned maneuvers at the altitude at which the airshow display is to be flown. The only safe way to acquire low-level aerobatic skills is to get thorough dual instruction from an experienced low-level display pilot.

It is a widely held opinion in the aerobatic community that in order to be an aerobatic display pilot it is best to first become a good competition pilot. The demanding standards of competition aerobatics gives you the fundamental skill set, confidence, and experience on which you can build the low altitude skills to become a safe airshow performer. A solid competition record will also give you the credibility you need to be offered the kind of financial compensation by the organizers that can make a reasonable contribution to your annual income. Given the aerobatic display standards seen at airshows today, you should be able to fly aerobatics at least at the national advanced category competition level to have any chance at getting steady airshow work.

When your general aerobatic skills are up to par, the biggest challenge is learning to fly safely at low level. Below approximately 600 feet, the perspective of the environment begins to change. Until you learn to accurately interpret its clues, you will not have a reliable sense of your position, and a small error can extract an immense price. The appearance of the horizon's position relative to the aircraft is different. Your perception of ground proximity might be misleading. Distances and rates of change in airspeed become more difficult to judge. You must readjust your sense of perception to be able to accurately judge your position relative to the low-level environment. There is

Fig. 19-3. *Awaiting your turn.*

nothing mysterious about it. It has to be learned from someone in the other seat who already knows and is qualified to teach it.

The other critical factor in low-level aerobatics is energy management. If you bust the floor of the competition box by 50 feet because you didn't regain sufficient height from your previous maneuver to have enough energy for the present maneuver, you can resolve not to allow it to happen again. If you bust the floor of your ground-level waiver you won't get another chance. The objective is to always have some reserve energy, a margin of error whether expressed in altitude or airspeed, as appropriate for

Patty Wagstaff Airshows

Fig. 19-4. *Three-time U.S. National Aerobatic Champion Patty Wagstaff delighting the air-show crowds.*

Pompano Air Center

Fig. 19-5. *Smoke's on . . . . Formation airshow aerobatics demand special skills.*

a particular stage of a maneuver. Know what the margins are for every single foot of your entire airshow sequence. A planned pullout at 200 feet from a hammerhead with a generously long vertical downline allows you to pull out sooner from a hammerhead gone wrong, giving you a big margin of error provided your sensory perceptions accurately tell you that you blew it and it is time to pull out right now.

Another important safety consideration in airshow aerobatics is to avoid on-the-spot sequence improvisation. Always fly a planned sequence and stick to it. If in the interest of safety a deviation is required, know in advance what your escape will be. Build a simple alternative component for each segment of your sequence that you can use to reposition yourself in a visually pleasing way if the maneuvering sequence is running short on energy. Be ready at any moment to completely abandon the sequence if you are the least bit uncomfortable with how it is going. Avoid any fixation on having to complete the show.

Hot and humid days and high field elevations will rob the airplane noticeably of performance. Somehow you must get the practice necessary to fly your sequence safely under these conditions. You could wait for the right circumstances, but a much easier and more productive way is to get your sequence nailed at 100 percent power as well as 80 percent power.

A big challenge of low-level aerobatics, if not properly planned for, is what to do if the fan stops. You should know for every foot of your sequence exactly what you are going to do in case of engine failure, and you should design the sequence to give you an option. If engine failure occurs in the mid to upper altitudes of your performance the options are better than they might first appear because usually you are over a big, inviting airport; however, if you are in the lower altitude segments of your sequence, you might face a tricky situation. Your prospects will be good in such a predicament if you are carrying a lot of energy and can pull up to a reasonable altitude—factors to bear in mind when composing your sequence.

An important consideration not only from the organizers' point of view but also from the pilot's perspective is the safety of the crowd. A concern hovering in the back of the general public's mind regarding airshows is an airplane falling on the spectators. As of this writing this has not happened at an American airshow in several decades. Flight line and crowd placement and a prohibition on flying over the crowd have had a lot to do with this excellent record, but that is just part of the story. Credit is also due to the pilots and how they plan their aerobatic performances. If a sequence is poorly planned, the aircraft could be in a position at certain points to jeopardize the crowd in the highly unlikely event that it went out of control. A well-planned sequence ensures that the crowd is never threatened, regardless of where control might be lost.

DEVELOPING YOUR ACT

The essence of airshow aerobatics is to please the crowd, to deliver what they want to see. For them, aerobatics is an airplane tumbling and twirling across the sky in a magical dance. The more a spectator can ask "How did they do that?!" the greater the pleasure. The typical airshow spectators have very little understanding of the finer points and complexity of aerobatics in its highest form, so they are quite easily pleased with much less than a no-holds-barred unlimited aerobatic competition sequence to medal winning standards.

It is a bit like comparing a commercial touring ice show to the world figure skating championships. The ice show skater is expected to know all the figures and perform them in a flashy, entertaining way that is not as exacting as the standards of top level skating competition. This "show style" is great for experienced competition pilots because, while requiring consummate skill, it also gives them a chance to go out there and really enjoy themselves instead of having to worry about being 5° off at the end of a spin recovery.

So, how to compose an airshow act? The details are a matter of individual preference, but there are some common rules of thumb worth considering. A cardinal rule is never to be straight and level. That is boring and entirely unexpected from an aerobatic airplane. The airplane should always be going up, down, at an angle, and preferably from one maneuver right into the other. A fast-paced high-energy sequence is far more exciting to most spectators than a slow lyrical sequence. Many noncompetition maneuvers will always thrill a crowd, such as the Lomcovak, torque rolls, and any other wild gyration where the airplane tumbles across the sky in an apparent defiance of gravity and without any visible means of support from the wings.

Rolls of all kinds are also popular, especially at high roll rates and with long roll sequences. An innovative smorgasbord of fairly simple maneuvers that have low difficulty coefficients in competition aerobatics can easily drive an airshow crowd wild, especially with a lot of multicolored smoke and a dynamic musical score in tune with the times. Airshow spectators are always on the lookout for something different, so an unusual airplane or a dynamite paint job might do a lot to distinguish you from all the Pitts acts.

An important practical consideration in designing an airshow sequence is the weather. While it obviously cannot be predicted, you can and should have contingency plans. If the ceiling is low, but VFR, and visibility is reasonable, the show usually goes on and you might have a problem flying an aerobatic sequence with vertical elements. For this reason most pilots have a high-altitude sequence and a low-altitude sequence and are ready to fly either one depending upon the weather.

A nice feature of airshow flying is that it provides an opportunity to explore forms of aerobatics not possible in competition flying. Re-creations of old barnstorming days, formation aerobatics, warbird aerobatics, and flying farmer antics are all huge crowd pleasers. They are also each highly specialized forms of aerobatics requiring thorough training from the right people and long hours of practice.

GETTING ON THE AIRSHOW CIRCUIT

If you are persistent and your act is good, sooner or later you will make it. A good act is key, but it means a lot more than the fine performance of a routine flown by everyone else. For an act to be in great demand, it has to be original. Either the flying has to be different from what's already out there or the airplane has to be exotic—preferably both. There was a time when—if you had a MiG 15—you would have been booked solid even if you could fly it only straight and level. If you can do six vertical rolls in a Sukhoi 26 followed by three practically stationary torque rolls hanging by the prop, you will find your services in great demand. The airshow pilot who has wheels on top of his Pitts and can land it upside down as well as right side up seems to be busy all the time.

When you have a good act, you have two choices. The low-budget option is to contact the local grass roots airshows and offer your services on their terms. As you gain experience and exposure, you can work your way up into the bigger shows, start getting larger fees for your appearances and eventually even make a little money. The high-cost option is to have a publicity package produced by a public relations or communications firm for a full-blown marketing campaign. If your act is reasonably good, the professional marketing approach might yield faster results compared to the grass roots approach.

A well-known aerobatic pilot with top national and international competition honors might be able to get a commercial sponsor for the airshow circuit. Some makers of mass market goods provide lucrative sponsorships.

Aerobatic teams have an especially good track record of obtaining sponsorships because of the greater exposure they can give the sponsor's products than a single pilot. Among the companies that have sponsored individual acts or formation teams are Budweiser, Pepsi, Mopar, Toyota, B.F. Goodrich Aerospace, and LapMap.

However you do it, once you are in, success is up to you. Keep polishing your act, strive always to innovate, be safe, and have a great time.

20
Buying an aerobatic aircraft

AFTER COMPLETING A BASIC AEROBATIC TRAINING PROGRAM YOU CAN rent an airplane like a Decathlon solo at most flight schools, and many pilots do. But if you hanker for aerobatics on your own in a Pitts, you are unlikely to find one available for solo rental. Nor will the rental option be able to meet your needs if you want to become a serious competition pilot or if you want greater flexibility in using an airplane. For any aerobatics beyond occasional basic maneuvers in a low-performance trainer, your only option in the United States is to buy an aerobatic airplane.

Ownership raises many questions: what aircraft to buy; what will it cost; should you consider a partnership; where to find aerobatic airplanes for sale; how to check them out reliably; what to think about homebuilts. The prospect of buying an aerobatic airplane can be a bit overwhelming, especially for first-time buyers, but if you break the process down into its components and take the time to understand a handful of fundamental guidelines, you will find the experience straightforward and trouble-free.

Buying a factory-new aircraft is as easy as buying a new car, when you already know the type you want and how you will pay for it. If you are buying a used airplane, which is what most of us on a budget are likely to do, the selection of the type is only the first step. You will have to find a specific airplane for sale at a reasonable price and you will have to thoroughly check out its mechanical condition.

WHAT AIRCRAFT TO BUY

The best rule of thumb is to get an airplane within the limits of your experience and suitable for the kind of flying you intend to do. Finances are also a consideration, but we will worry about that later. You should first choose an airplane you really want and then see if you can find a way to make the finances work.

If your primary purpose is recreational aerobatics, consider the aerobatic use and other potential uses of the airplane. From an aerobatic standpoint, your big decision will be whether or not to get an airplane with an inverted system. Without an inverted system, you are restricted to positive-G maneuvers, which is fine if all you want to do is aileron rolls, inside loops, and spins. Nonaerobatic considerations might include cross-country capabilities, cabin comfort, avionics options, and passenger capacity.

If aerobatics is your main objective, you will most likely want an airplane built for the purpose, with nonaerobatic characteristics being a secondary consideration. Your choice will be a matter of personal preference and finances, but in the interest of safety, you should avoid getting too much airplane too soon. It is best to work up to the unlimited machines gradually. Get an airplane only if you can fly aerobatics in it well enough to be safe. Trade it in for a more advanced machine only when it no longer has anything to teach you. Buying an airplane that is capable of far more than you are is not only a waste of resources, but it can also be counterproductive. You might end up being afraid of it and your confidence could suffer.

Important questions are whether to buy a new or a used airplane (if the airplane is still in production), and if you should consider homebuilts. Usually, a used airplane is much less expensive than a new one. While it is dangerous to generalize because of wear and tear and supply and demand, aircraft do seem to lose a greater percentage of their value in the first few years over and above the use they are subjected to, after which the price stabilizes and becomes less a matter of age and more a function of hours flown and physical condition. Unless there is a design change that affects flying characteristics, an impeccably maintained used airplane with an engine recently overhauled by a reputable shop to factory new tolerances is the best deal financially. Beyond the engine, considering the potential hard use of an aerobatic airplane's airframe, it is important to be able to get a reasonable idea of how many hours it was flown aerobatically, how hard it was flown, and how much stress it might have been subjected to.

Some of the most outstanding aerobatic airplanes are homebuilts, and in principle they can be as good a choice as any other option; however, it is much more difficult to ascertain the quality of construction in a homebuilt than it is in a production airplane. In addition to gathering the usual used airplane facts, if you are considering a homebuilt, it is imperative to develop reliable information on the builder's skills, experience, and reputation, and the quality of work that went into the particular airplane you are looking at.

Unless you are a skilled A&P, plan to use an experienced mechanic familiar with aerobatic aircraft to do the prepurchase inspection. We will discuss what to look for during the prepurchase inspection later on in this chapter, but first let's get some ideas on where to find aerobatic aircraft for sale and what they will cost to buy and fly.

Chuck Stewart

Fig. 20-1. *The classic Zlin Trener is an excellent buy for the recreational aerobatic pilot.*

WHERE ARE THEY?

Factory new airplanes are easy to find. Just call the nearest dealer. Most dealers advertise in the International Aerobatic Club's monthly magazine, *Sport Aerobatics*. If you have a problem finding a dealer or a factory address for a particular type of airplane that you think is in production, call the IAC for assistance.

A variety of good information sources on used and homebuilt airplanes are available. You should consult as many of them as possible to be well informed about the market, but you might feel most comfortable if you can find a particular airplane you know well, or one that is recommended by a knowledgeable person in the field of aerobatics whose judgment you trust. The network of pilots and mechanics in your local IAC chapter, at your aerobatic flight school, and in the aerobatic community at large are all excellent sources of information.

Dealers of aerobatic aircraft also handle used aircraft, and many of them advertise this service in *Sport Aerobatics*. Often they have an intimate knowledge of a used airplane they are brokering, having originally sold it to the present owner.

Aviation publications are another good source of information, especially the classified section of *Sport Aerobatics* and the aerobatic section of *Trade-A-Plane*. If you find a used production airplane or a homebuilt that sounds promising in the print of an advertisement, leave no stone unturned to verify its background through the network of the aerobatic community. Let the buyer beware!

WHAT IT WILL COST TO BUY AND FLY

The reasonable acquisition cost of an aerobatic aircraft is easy to find out, but do your homework to avoid overpaying. If you are buying a new airplane from a dealer, do not be reluctant to bargain as you would with any car dealer. Contact several dealers, if there are more than one, for the type you are considering. Theoretically, dealers of new aircraft might have exclusive territories for airplanes with small production runs, but with a bit of planning you can always find a plausible reason to purchase through the territory where you find the best price.

If you are buying a used aircraft or a homebuilt, review and compare all the price information in the advertisements in the various aviation publications and consult knowledgeable friends, acquaintances, and friends of friends in the aerobatic community. You will soon have an idea of the going rate. Bear in mind that advertised prices are rarely final. There is plenty of room to horse trade.

Fig. 20-2. *If the price is too high, consider a partnership.*

Aviat

Fig. 20-3. *There is nothing like the first loop after you forked over the check.*

If the aircraft is listed in the *Aircraft Blue Book Price Digest* try to find out what the typical values are. (The digest is available only to brokers and aircraft financiers who are not supposed to pass on the information to the general public.) A good way to do it is to give a banker who specializes in aircraft financing the specifics on an advertised aircraft that sounds good, and ask what would be the maximum financing available to a qualified buyer. The typical aviation bank finances maximum 80 percent of the retail blue book value of an airplane, so you can easily figure out the full retail price based on the bank's quote.

Most people have no trouble finding out the reasonable purchase price of an aircraft. Where they are more likely to encounter difficulties is in figuring out what the airplane would cost to own and operate after they bought it. The costs are easy to estimate if you follow a few simple guidelines.

We will examine the single owner's costs of operating a trainer (such as a Decathlon), an intermediate aircraft (such as a Pitts S2-B), or an unlimited aircraft (such as a Sukhoi Su-29, an Extra 300, or a Staudacher 300). While taken from real examples at the time of writing, the figures used are indicative. Prices and costs can and will fluctuate wildly depending upon a variety of complex factors. The intention is not to provide specific cost information, but to familiarize you with the method of estimating the costs of aerobatic flying.

Let's briefly review the cost components of aircraft ownership. The two basic cost categories are *fixed expenses* and *operating expenses*. Fixed expenses are expenses that you have to pay regardless of the hours flown by the airplane (such as loan payments, insurance, and state fees); fixed expenses are commonly measured annually.

Operating expenses are expenses incurred directly as a result of operating an airplane (examples are fuel and oil); operating expenses are commonly measured hourly.

Fixed and operating expenses are combined to measure total flying expenses two ways: total hourly and total annual:

- Total hourly expenses are annual fixed expenses divided by the hours flown per year (to get hourly fixed expenses), plus hourly operating expenses.
- Total annual expenses are total hourly expenses times the number of hours flown per year.

Expenses

Fixed expenses and operating expenses are each composed of a handful of easy-to-understand expense categories.

Annual fixed expenses

- Tiedown/hangar. The cost of aircraft storage.
- Insurance. The annual insurance bill. It can vary widely from insurer to insurer and depends on the extent of coverage.
- State fees. A collective category for various state fees and taxes for the year. It will vary from state to state.
- Annual. The cost of the annual inspection only. The flat fee charged to inspect the airplane. Any repairs are extra.
- Maintenance. There are two schools of thought on the appropriate category for maintenance expense. One holds that maintenance is strictly an operating expense, not a fixed expense, to be reserved against entirely per flying hour. The other school holds that some maintenance has to be done every year, regardless of whether or not the airplane flies. The latter school believes that some "fixed" maintenance should be estimated every year, in addition to maintenance covered by the general maintenance reserve under hourly operating expenses. This category provides for an annual maintenance amount in addition to the hourly reserve. It is up to the user to decide where to reserve. To be conservative, it is best to reserve an annual fixed amount for maintenance in addition to an hourly reserve.
- Loan payments. The annual payments on any loan taken out to finance the aircraft.
- Cost of capital (noncash). The opportunity cost of capital. The amount per year lost in income on the capital used to purchase the airplane instead of being invested in an income earning investment. It is not a cash expense because it is not paid out of pocket. Rather, it is money that was not put in your pocket to begin with, but would have been, had you left the capital in the bank instead of buying an airplane with it.

	Trainer	Intermed.	Unlimited
NUMBER OF PILOTS	1	1	1
ANNUAL FIXED EXPENSES			
Tiedown/Hangar	3000.00	3000.00	3000.00
Insurance	1800.00	2800.00	4000.00
State Fees	150.00	150.00	150.00
Annual	500.00	500.00	800.00
Maintenance	1000.00	1000.00	1200.00
Loan Payments	3568.07	6739.69	11893.57
Cost of Capital (non-cash)	1575.00	2975.00	5250.00
Total Fixed Expenses / yr	11593.07	17164.69	26293.57
HOURLY OPERATING EXPENSES			
Fuel	21.00	42.00	52.50
Oil	0.80	1.60	6.40
Engine Reserve	8.00	12.00	16.67
General Maint Res	10.00	10.00	15.00
Total Op Exp / hr	39.80	65.60	90.57
TOTAL HOURLY EXPENSES			
50 Hours	271.66	404.89	616.44
100 Hours	155.73	237.25	353.50
Hourly Commercial Rental	100.00	195.00	0.00
TOTAL ANNUAL EXPENSES			
50 Hours, Own	13583.07	20444.69	30821.90
Own (cash only)	12008.07	17469.69	25571.90
Rent	5000.00	9750.00	0.00
100 Hours, Own	15573.07	23724.69	35350.23
Own (cash only)	13998.07	20749.69	30100.23
Rent	10000.00	19500.00	0.00

Hourly operating expenses

- Fuel. Per hour cost of fuel ($*gal/hr).

- Oil. Per hour cost of oil ($*qt/hr).

- Engine reserve. The money set aside for each hour of flying to build up a reserve specifically earmarked for overhauling the engine. It is a cash expense because it should be paid into a bank account where it accumulates until overhaul time. It is not intended for any engine maintenance other than overhaul.

- General maintenance reserve. Similar to the engine reserve, but it is intended to ensure the availability of funds for general maintenance. Do not confuse this reserve with the fixed maintenance expenses that cover an estimated amount of maintenance required every year regardless of whether or not the airplane flies. Many pilots use this reserve to meet all routine maintenance expenses instead of reserving in part under annual fixed expenses.

ASSUMPTIONS: (\$ unless shown otherwise)	Trainer	Intermed.	Unlimited
Aircraft Value:	45,000	85,000	150,000
Loan Amount	22,500	42,500	75,000
Hangar/month:	250	250	250
State fees:	150	150	150
Insurance/yr:	1,800	1,500	2,200
Annual:	500	750	1,000
Maintenance/yr:	1000	1,500	3,000
Fuel Cost (\$/gal):	2.10	2.10	2.10
Oil Cost (\$/qt):	3.20	3.20	3.20
Fuel Cons (gal/hr):	10	20	25
Oil Cons (qt/hr):	0.25	0.50	2.00
Engine Reserve/hr:	8.00	12.00	16.67
Gen Maint Res/hr:	10.00	10.00	15.00
Time Before OH (hrs):	1,500	1,500	1,200
Engine MOH Cost:	12,000	18,000	20,000
Loan Interest Rate:	10%	10%	10%
Loan Maturity:	10 yrs	10 yrs	10 yrs
Cost of Capital Rate:	7 %	7 %	7 %

All you have to do to estimate the costs of flying any aircraft is to gather accurate data for all the cost components and summarize them. Here is what the data for a typical used trainer, intermediate aircraft, and unlimited aircraft looked like at the time of writing. Note that in all cases, 50 percent of the purchase price is borrowed. Financing is expensive. Without financing, fixed expenses would be significantly lower. Total annual expenses are shown for 50 hours and 100 hours of annual flying time. For different annual hours, multiply hourly operating expenses by the number of hours you plan to fly per year.

Expensive, isn't it? Do not despair because there is a less expensive alternative in the form of an aircraft partnership.

THE PARTNERSHIP OPTION

Aircraft partnerships can dramatically cut the per person cost of flying. From the standpoint of aircraft availability, partnerships are ideal for aerobatic aircraft that are used intensively but are flown for relatively short periods of time per flight. In one day, two partners can easily get in three aerobatic practice sessions apiece, which is about all the human body can stand.

The weak point of an aircraft partnership is the human factor. It is essential that you and your partners be absolutely compatible in terms of how you fly and care for the aircraft, otherwise the partnership will break down.

A smooth working partnership has to be structured well, as spelled out in an effective partnership agreement. It is beyond the scope of this book to go into further details on setting up and operating a partnership, but you are encouraged to read *Fly For Less, Flying Clubs and Aircraft Partnerships*, by Geza Szurovy, TAB Books, a division of McGraw Hill, Inc., 1992.

To illustrate how dramatically partnerships can cut the per-person cost of flying, look at table on facing page. It compares the costs of a single owner to partnerships of two and four pilots for a trainer, an intermediate aircraft, and an unlimited aircraft. For a breakdown of the numbers with more partners, see Appendix D, which presents each aircraft's costs for up to 15 pilots. Bear in mind that beyond five partners most insurance companies require the joint owners to be technically organized as a flying club.

Some people have such exacting personalities that for them a partnership is not an alternative, and if you are good enough to be among the handful of top national unlimited competition pilots you will probably want to find a way to own alone for maximum flexibility to fully concentrate on your flying. But for most of us mortals, a partnership might well be the key to enjoying what we otherwise couldn't afford.

CHECKING OUT AND BUYING THE AIRCRAFT

Having decided what aircraft type to buy, sorted out the anticipated cost, and located specific aircraft available for sale, it is time to check them out and buy one. If you are lucky enough to be buying a new aircraft, get together with the dealer or the factory, and you are on your way. If you are buying a used aircraft or any homebuilt, settle down for some further homework.

Initial contact with the seller

If you are well-organized and ask the right questions when you make the initial phone call, you can quickly screen available aircraft to decide which ones merit further follow-up. Make up a blank questionnaire and record all the answers for reference. Typical questions to ask are:

- How many owners has the airplane had?
- Where has it been based geographically during its life?
- What type of aerobatic flying has the aircraft done?
- What contests has it flown? (Get the whole record and check it.)
- Total time, airframe?
- Total aerobatic time, airframe?
- Total time engine?
- Total aerobatic time, engine?
- Total time engine since major overhaul?
- How many times has the engine been overhauled?
- Who performed the overhaul?

NUMBER OF PILOTS	Trainer			Intermediate			Unlimited		
	1	2	4	1	2	4	1	2	4
ANNUAL FIXED EXPENSES									
Tiedown/Hangar	3000.00	1500.00	750.00	3000.00	1500.00	750.00	3000.00	1500.00	750.00
Insurance	1800.00	900.00	450.00	2800.00	1400.00	700.00	4000.00	2000.00	1000.00
State Fees	150.00	75.00	37.50	150.00	75.00	37.50	150.00	75.00	37.50
Annual	500.00	250.00	125.00	500.00	250.00	125.00	800.00	400.00	200.00
Maintenance	1000.00	500.00	250.00	1000.00	500.00	250.00	1200.00	600.00	300.00
Loan Payments	3568.07	1784.03	892.02	6739.69	3369.84	1684.92	11893.57	5946.78	2973.39
Cost of Capital (non-cash)	1575.00	787.50	393.75	2975.00	1487.50	743.75	5250.00	2625.00	1312.50
Total Fixed Expenses / yr	11593.07	5796.53	2898.27	17164.69	8582.34	4291.17	26293.57	13146.78	6573.39
HOURLY OPERATING EXP.									
Fuel	21.00	21.00	21.00	42.00	42.00	42.00	52.50	52.50	52.50
Oil	0.80	0.80	0.80	1.60	1.60	1.60	6.40	6.40	6.40
Engine Reserve	8.00	8.00	8.00	12.00	12.00	12.00	16.67	16.67	16.67
General Maint Res	10.00	10.00	10.00	10.00	10.00	10.00	15.00	15.00	15.00
Total Op Exp / hr	39.80	39.80	39.80	65.60	65.60	65.60	90.57	90.57	90.57
TOTAL HOURLY EXPENSES									
50 Hours	271.66	155.73	97.77	408.89	237.25	151.42	616.44	353.50	222.03
100 Hours	155.73	97.77	68.78	237.25	151.42	108.51	353.50	222.03	156.30
Hourly Commercial Rental	100.00	100.00	100.00	195.00	195.00	195.00	0.00	0.00	0.00
TOTAL ANNUAL EXPENSES									
50 Hours, Own	13583.07	7786.53	4888.27	20444.69	11862.34	7571.17	30821.90	17675.12	11101.72
Own (cash only)	12008.07	6999.03	4494.52	17469.69	10374.84	6827.42	25571.90	15050.12	9789.22
Rent	5000.00	5000.00	5000.00	9750.00	9750.00	9750.00	0.00	0.00	0.00
100 Hours, Own	15573.07	9776.53	6878.27	23724.69	15142.34	10851.17	35350.23	22203.45	15630.06
Own (cash only)	13998.07	8989.03	6484.52	20749.69	13654.84	10107.42	30100.23	19578.45	14317.56
Rent	10000.00	10000.00	10000.00	19500.00	19500.00	19500.00	0.00	0.00	0.00

ASSUMPTIONS:	Trainer	Intermed.	Advanced
($ unless shown otherwise)			
Aircraft Value:	45,000	85,000	150,000
Loan Amount	22,500	42,500	75,000
Hangar/month:	250	250	250
State fees:	150	150	150
Insurance/yr:	1,800	1,500	2,200
Annual:	500	750	1,000
Maintenance/yr:	1000	1,500	3,000
Fuel Cost ($/gal):	2.10	2.10	2.10
Oil Cost ($/qt):	3.20	3.20	3.20
Fuel Cons (gal/hr):	10	20	25
Oil Cons (qt/hr):	0.25	0.50	2.00
Engine Reserve/hr:	8.00	12.00	16.67
Gen Maint Res/hr:	10.00	10.00	15.00
Time Before OH (hrs):	1,500	1,500	1,200
Engine MOH Cost:	12,000	18,000	20,000
Loan Interest Rate:	10%	10%	10%
Loan Maturity:	10 yrs	10 yrs	10 yrs
Cost of Capital Rate:	7%	7%	7%

~ The choices are the factory (the best but most expensive alternative), an FAA-certified repair station (many specialize in engine overhauls and are perhaps the best buy for the money), or an FBO or independent mechanic (who do some of the best and some of the worst work around; the problem is knowing who does the best and who does the worst).

~ Was the overhaul to *factory new tolerances* or *service limits*? Factory new tolerances mean that the main components meet the standards of new components. Service limits mean that wear and tear is within the limits within which the components can be kept in service, but the components might be considerably below factory new tolerances.

~ Were the accessories also overhauled? If not, how many hours are on the accessories? (Starter, alternator, magnetos, vacuum pump, etc.)

• What does the aircraft weigh? Excess weight caused by optional accessories can diminish the performance of some aerobatic aircraft.

• Are all ADs complied with and entered into the aircraft logs?

• Has there been a top overhaul since the major overhaul?

• Is there any damage history?

• Are the flying wires, if any, in good condition?

• How old is the paint and interior?

• Rate the exterior and interior on a scale of 1 to 10.

• Has the airplane been hangared?

- When was the last annual?
- Who did the last annual?
- How often is the oil changed?
- Is the oil sent out for analysis at oil change? Are the results available?
- How many hours has the airplane flown since the last annual?
- How many hours has the airplane flown in the last 12 months?
- What are the major maintenance items from the last 12 months?
- What make and model avionics does the airplane have?
- Are there any maintenance issues with the avionics?
- Does the airplane have an intercom? Does it work?
- Are the airframe and engine logs complete?
- Does a parachute come with the airplane and what kind is it?
- Are pictures available? (exterior, interior, and instrument panel)
- What is the asking price?

A homebuilt airplane purchase requires additional questions:

- When was the aircraft built?
- Who built it?
- Can the builder be reached? (talk to the builder in great detail)
- What experience did the builder have?
- Has the builder built any other aircraft? Can the owners of those aircraft be contacted?
- Was the aircraft kit built or scratch built?
- What components, if any, were prebuilt by a professional company?
- What records are there of the building process? Bills, pictures, videotapes, and the like? Are the FAA sign offs readily available?

Your follow-up decision might be influenced by such factors as total aerobatic time, total time, maintenance practices, and ease of getting together with the seller for the prepurchase inspection.

The prepurchase inspection

A thorough prepurchase inspection with your own mechanic is at the heart of making a decision about buying a specific aircraft. If you are buying a homebuilt, arrange for someone very knowledgeable about the type to look at it with you, or at least talk to your mechanic on the phone. Gather information from type associations about what weaknesses and quirks to watch for.

A prepurchase inspection usually consist of three phases:

- Examination of the aircraft papers
- The mechanical inspection of the airplane
- The flight test

Use these guidelines and rely on your mechanic for best results.

Fig. 20-4. *The custom-built construction of a Staudacher 300GS for Mike Goulian.*

Aircraft papers

If the aircraft papers don't check out, there is no reason to examine the aircraft, so you can save yourself a lot of time if you start with the paperwork. Concentrate on the airframe and engine logs, but examine all papers required by the FAA.

- Certificate of airworthiness
- Registration
- Radio station license
- Airframe and engine logs

Check for:
Proper annual entries
Evidence of 100-hour inspections indicating commercial use
Compression (most recent and history)
Amount of hours flown per year
Evidence of unscheduled repair work
Airworthiness directive compliance
Engine major overhaul entry (when and where done, to what tolerances)
Record of geographic movements

Weight and balance

Check for excessive empty weight

MECHANICAL INSPECTION

The mechanical inspection should be along the lines of a 100-hour inspection. Given the limits to which aerobatic aircraft are flown, the airframe inspection should be extremely thorough and your mechanic should know exactly what signs of aerobatic wear to look for beyond regular wear and tear. Regarding the engine, it is especially important to do a compression check and to verify overhaul information. Consider these guidelines for a general idea of what to inspect.

Airframe

Check for wrinkled skin, loose rivets, dings, cracks, corrosion, and signs of skin or fabric separation.

Check for mismatched paint, which could be a sign of repairs.

Check all controls for free and correct movement.

Check all control hinges (ailerons, elevator, rudder, flaps) for looseness, play, and hairline cracks.

Check vertical and horizontal stabilizer attach points for looseness, play, and hairline cracks.

Check wing attach points for hairline cracks and corrosion. (Looseness equates to immediate rejection of the aircraft.)

Check control cables for looseness and chafing. Look inside the fuselage and wings through inspection panels with a flashlight.

Check the fuselage interior for any loose items.

Check fuel caps, quick drains and fuel tank areas for signs of fuel leaks (brownish stains).

Check fuselage underside for cleanliness and signs of leaks from engine area.

Check wing struts for any signs of damage, corrosion, or hairline cracks.

Check the flying wires, if any, for nicks and cracks.

On fabric-covered aircraft, do a fabric test and check for loose or peeling fabric.

Check engine cowling for looseness, play, and cracks, especially at attach points.

Landing gear

Check landing gear struts for leaks.

Check tires for wear and bald spots.

Check brakepads for wear, brake disks for corrosion, pitting, and warping, and brake hydraulic lines for signs of seeping or leaking fluid.

Cockpit

Check canopy or cabin doors and windows for signs of water leaks.

Check windows for crazing.

Check regular seat belts and the aerobatic harness for wear and tear.

Move all controls and trim to verify full control movement and check for binding.

Observe G meter reading and check for proper operation.

Move all other knobs and switches to check for proper operation. Check ELT for proper operation.

Check entire aircraft for proper display of placards and limitations.

Engine and propeller

Compression check.

Check baffles for damage or deformation. Baffle irregularities can cause cooling problems.

Check for any sign of leaks, especially around the various gaskets. Look for oil and fuel stains.

Check lower spark plugs for proper condition (take them out and examine them).

Check wiring harness for signs of brittleness and fraying.

Check the induction/exhaust system for leaks, cracks, corrosion, and looseness.

Check engine controls running to the cockpit for free and easy movement.

Check the battery for fluid level and signs of overheating.

Check accessory attachments (alternator, mags, vacuum pump, starter, electric fuel pump, etc.). Check alternator belt for fraying and tightness.

Check propeller for nicks, and spinner for cracks.

TEST FLIGHT

The test flight is a delicate matter in an aerobatic aircraft. It is one thing to go for a ride around the patch in a run-of-the-mill multiseat light airplane and quite another to jump into an aerobatic airplane you don't really know and fly it to its limits either with an owner you also don't know, if it is a two-seater, or alone if it is a single-seater (in the latter case insurance terms might also restrict you from flying the airplane). Yet somehow you must find out what the airplane is capable of. A good way to do it is to arrange an aerobatic demonstration flight by the owner and observe it from the ground. You can collaborate on the ground portion of the demonstration (engine start, taxiing around, etc.) and you can check the G meter when the owner returns to verify the loads put on the airplane during the demonstration.

You and the owner

Preflight. Do a detailed preflight per the operating manual. Follow the manual's detailed checklist to be sure of covering everything. Set G meter to zero.

Engine start. Note how easily the engine cranks and starts.

Taxi. Perform radio checks. Test brakes. Notice the engine response, steering, suspension.

Pretakeoff checks. Perform careful pretakeoff checks and runup verifying aircraft behavior to operating manual standards.

Owner alone

Flight demonstration. Have the owner perform an aerobatic demonstration, preferably at least a Sportsman sequence, at a safe altitude and observe it carefully. Not only will you see the aircraft flying the maneuvers, but you will also be able to tell a lot about how it might have been used, based upon the owner's performance.

If the aircraft is a two-seater, following the owner's satisfactory solo demonstration you might want to go up with the owner and fly the sequence yourself. In this case take the opportunity to make sure everything works on the airplane as advertised. If the aircraft is a single-seater, insurance considerations and the owner's lack of familiarity with your flying skills might prevent you from flying it until you buy it.

You and the owner

Postflight check. Check the G meter. Uncowl the engine, check for evidence of any leaks. Turn on all navigation lights, landing lights, strobes, and flashing beacon, and check for operation.

If everything checks out, you are in business. Buy the airplane—the bank will guide you through the paperwork if you are financing the deal, or ask your local FSDO for guidance—and go flying. Work up to the more complex aerobatic maneuvers slowly at a conservative safety altitude until you are fully comfortable with the airplane, and enjoy it up there . . . .

Stearman sortie

There is no escaping that aerobatics is energy management, and nowhere is the scarcity of energy available for aerobatics more apparent than in the grand old biplanes of yesteryear. A stock Stearman takes off at 70 knots, cruises at 70 knots, and lands at 70 knots. Like most biplanes of its generation and before, it has a huge drag factory of an airframe, fat asymmetric wings, a truly lazy roll rate, and much too little power by modern standards. Compared to the Stearman, the little training Decathlon is a lean, lightning-quick cougar. (Hanging a 500-hp engine on the Stearman will, of course, cure all, but that's cheating.)

Aerobatics takes on a graceful, serene meaning of its own in an antique biplane, the stuff of silk scarves and sunny summer afternoons. And learning to do aerobatics well in a Stearman will make you a much better pilot. The minimal energy reserve puts you closer to the airplane's limits in even the most simple maneuvers, and the consequences of any errors are magnified in comparison to the more modern machines. If you have enough wrist action to stir your coffee, you can roll a Pitts. Manhandled controls in the Decathlon will make a slow roll sloppy and you will lose some altitude.

But if you run out of energy as you roll inverted in a Stearman, stand by for one thrilling elevator ride. Down. By the time you properly learn to slow roll an antique biplane, you will have explored and come to understand every nook and cranny of the maneuver as in no other aerobatic airplane, and the same holds true for most other maneuvers. And you will have discovered a new sensory instrument: your cheeks and the wind striking them in a skid or a slip.

It is all true about the old biplanes: the silk scarves, the oneness with the elements, ghostly wingmen appearing from sunny summer afternoons long gone, the pleasure of a roll done well. Indulge before it all disappears into the history books.

21
Aerobatic aircraft
type directory

A WIDE RANGE OF NEW AND USED AEROBATIC AIRCRAFT ARE AVAILABLE
to satisfy every aerobatic objective, piloting skill, and performance demand.
Many are in series production, some are no longer made, some are homebuilts, and
some are one of a kind that are built to a particular competitor's personal specifica-
tions. The airplanes range from antique biplanes to the most dynamic competition aircraft of the day. This chapter is not a compilation of all aerobatic aircraft that had a role
at some time or other in the story of aerobatics. Rather, it presents a sampling of his-
torically important machines and those aerobatic aircraft that are readily available to
the aerobatic pilot today. There is a suitable type for everyone, from the novice to the
unlimited competition pilot.

Pitts Special S-1 Series (S-1C, S-1S, S-1T)

The Pitts Special is unquestionably the most famous American aerobatic aircraft.
This tiny, overpowered biplane started out as a toy for its designer, Curtiss Pitts, in the
early 1940s. When American aerobatics experienced one of its periodic revivals during the 1960s, the Pitts Special was ready and waiting as the ideal competition machine. Its crisp handling and high power loading made it suitable for the dynamic style
of aerobatics then being popularized by the landmark Zlin Trener series of Czech
monoplanes. The Pitts was in a class of its own from the start in national competition
and by the late 1960s, in the hands of some of the world's best pilots, variants of the
Pitts Special were giving the Zlins a real run for their money internationally. Popularized by former women's world aerobatic champion Betty Skelton and the Red Devils
formation aerobatics team (now performing in Christen Eagles), the Pitts Special has

Type	Max. Speed (kt)	Climb Rate (ft/min)	Gross Weight (lbs)	Horse-power	Wing loading (lbs/sq ft)	Power loading (lbs/hp)	Limit Load (G)	Seats
Beech F-33 Bonanza	182	1,167	3,400	285	18.80	11.93		2
Beech T-34	167	1,100	2,950	225	16.60	13.11		2
Bücker Jungmann	102	630	1,474	100	9.49	14.74		2
Bücker Jungmeister	119	1,171	1,290	160	10.00	8.06		1
Cessna Aerobat	109	715	1,670	110	10.50	15.18		2
Christen Eagle	184	2,100	1,578	200	12.60	7.89	+7/-5	2
Citabria	140	725	1,650	150	10.00	11.00		2
Decathlon	160	1,000	1,800	180	10.60	10.00	+6/-3	2
DH Chipmunk	123	800	2,000	145	10.60	13.79		2
DH Tiger Moth	97	673	1,825	130	7.60	14.04		2
Extra 230	220	2,950	1,234	200	13.00	6.17	+10/-10	1
Extra 300	220	3,300	1,918	300	18.19	6.39	+10/-10	2
Great Lakes	114	1,150	1,800	180	9.36	10.00		2
Marchetti SF260	187	1,791	2,645	260	22.40	10.17	+6/-3	2
Mudry CAP 10	183	1,180	1,675	180	14.38	9.31	+6/-4.5	2
Mudry CAP 20	162	2,755	1,433	200	12.72	7.17	+8/-4.5	1
Mudry CAP 21	172	3,000	1,323	200	13.36	6.62	+8/-4.5	1
Mudry CAP 231	216	3,150	1,609	300	18.19	5.36	+10/-10	1
Pitts S-1S	176	2,600	1,150	180	11.70	6.39	+9/-4.5	1
Pitts S-1T	180	2,800	1,150	200	11.70	5.75	+9/-4.5	1
Pitts S-2A	176	1,900	1,500	200	12.00	7.50	+9/-4.5	2
Pitts S-2B	187	2,700	1,626	260	13.00	6.25	+9/-4.5	2
Slingsby Firefly	180	1,100	2,100	160	15.44	13.13		2
Spinks Acromaster	243	3,000	1,775	200	13.85	8.88	+8/- 8	1
Stampe SV-4	113	750	1,716	140	8.80	12.26		2
Staudacher S300	267	3,300	1,700	300	15.45	5.00	+12/-12	1
Staudacher 300GS	267	3,300	1,700	320	15.77	5.31	+12/-12	1
Stearman	110	840	2,717	220	10.94	12.35		2
Steen Skybolt	126	2,500	1,650	180	10.85	9.17		2
Stephens Akro	191	4,000	1,200	180	12.00	6.67	+12/-12	1
Sukhoi 26	243	3,540	2,205	360	17.36	6.13	+12/-10	1
Sukhoi 29	243	3,150	2,425	300	18.44	8.08	+12/-9	2
Yak 18	159	1,690	2,100	260	16.40	8.08	+9/-6	2
Yak 50	226	3,150	1,984	360	12.29	5.51	+ 9/-6	1
Yak 52	194	1,970	2,844	360	17.61	7.90	+9/-6	2
Yak 55	178	3,150	1,885	360	13.68	5.24	+9/-9	1
Zlin 50	185	2,360	1,582	260	11.80	6.08	+9/-6	1
Zlin 526	165	1,181	2,072	180	12.50	11.51	+7/-4.5	2

WARNING! These figures are illustrative strictly of approximate performance. Refer to the respective flight manuals for accurate specific data. Note:

1. Some speed figures are never exceed speed, some are maximum sea-level speed.
2. Limit load information is provided where available. U.S. aerobatic category certification is to +6/-3 G. Not all aircraft above are certified in the U.S. aerobatic category. See the appropriate aircraft documents for specific information.
3. The four seat Beech F-33 Bonanza may only carry two occupants on aerobatic flights; no occupant may be in the rear seats.

gone through many refinements and still remains competitive in the right hands in all categories, though increasingly outclassed for the top honors at the unlimited level by the new generation of monoplanes. Pitts aircraft are manufactured by Aviat.

For many years Pitts Specials were built by homebuilders. Plans are no longer available, but the total Pitts fleets is a good mix of homebuilt and factory built aircraft.

Pitts Special S-2 Series (S-2A and S-2B)

The S-2 series is the two seater version of the famous Pitts Specials. It was one of the first aircraft capable of flying unlimited aerobatics with two on board, and as such, is an excellent advanced aerobatic trainer. It is still in production by Aviat, the most recent purchaser of the Pitts production line and type certificate. The S-2 series is also popular with pilots who are recreational practitioners of advanced aerobatics and want the option of sharing the pleasure with a friend from time to time.

Christen Eagle

The Christen Eagle is the brainchild of a former owner of the Pitts factory, Frank Christensen. This two-seat sport biplane was designed in the late seventies and has gone on to become one of the most popular kit-built aircraft ever made. In addition to the aircraft's stellar performance, the superb quality of the kits has been an important factor in the type's popularity. Christensen took the utilitarian Pitts as his point of departure, and sought to refine it to the tastes of the recreational owner who might wish to do more on weekends than solely aerobatics. More cockpit creature comforts and better visibility make the Christen Eagle the choice for pilots who wish to travel in comparative comfort, yet still be able to perform aerobatics to Pitts Special standards. The Christen Eagle is available from Aviat.

Bellanca Decathlon

The Bellanca Decathlon's light, and pleasantly harmonious controls and benign flying characteristics make it an excellent basic aerobatic trainer. Though it is a descendant of the company's Citabria and Scout aircraft, the resemblance is largely superficial. It has a different airfoil and control system that give it superior aerobatic qualities. Since it was first built in the 1970s, this high-wing, fabric-covered aerobatic trainer has gone in and out of production several times and has had its share of ADs, especially the early models with wooden wing spars. As of this writing it is again in production, with metal wing spars and strengthened seats, and its gentle manners are likely to delight fledgling aerobatic students for years to come. It is also quite a comfortable cross-country machine.

Bellanca Citabria

Spell Citabria backward and you will get the idea. Regrettably the aerobatic handling characteristics of this airplane are about as much off the mark as the resemblance of its name to the word "aerobatic," when read backward: almost, but not quite there. Its flat airfoil and the early models' low horsepower have made it an outstanding utility aircraft, but not a particularly noteworthy aerobatic performer in comparison to the alternatives. Unless your idea of working out is frequent and vigorous arm wrestling, pass up the excessive stick forces generated by the Citabria in aerobatic flight, in favor of its more refined descendant, the Decathlon. (Spades that should lighten stick forces are available from a specialty supplier.)

Mudry CAP 10/20

The side-by-side two-seat French CAP 10 was developed in the late 1960s as a basic trainer for the French military. The French have always liked to work with wood when building light aircraft, and the all-wood CAP 10 was no exception. It quickly gained favor as a snappy basic and advanced aerobatic trainer as well as an all-around sport aircraft, and became quite popular with civilian pilots throughout Europe. In the United States, it is best known as the aircraft in which the husband-and-wife team in "The French Connection" put on their lyrical aerobatic airshow duet.

The CAP 20 is the single-seat version of the CAP 10. The only difference besides the single seat is a more powerful engine equipped with a constant-speed propeller.

Mudry CAP 21

The CAP 21 was Mudry's response to the Zlin 50 in the early 1980s. Superficially looking like a CAP 20, the 21 featured an all-new wing with an airfoil designed especially for aerobatics by Aerospatiale. Exceptionally low drag characteristics and ailerons spanning almost the entire wing gave the CAP 21 excellent penetration and an outstanding roll rate.

Mudry CAP 231

The Mudry CAP 231 was designed for unlimited aerobatic competition flying. It emerged in the late 1980s to challenge the more advanced German and Russian aircraft dominating the world competition arena. While Mudry utilized its experience gained on the earlier models, the CAP 231 was a completely new design. Its carefully chosen airfoil and high power loading rank it among the top aerobatic competition aircraft today.

Cessna Aerobat

Cessna produced a significantly beefed up version of the venerable 150 and dubbed it the Aerobat. Although the two aircraft look identical, the Aerobat's airframe is much stronger. In addition to strengthening elements incorporated throughout its structure, the Aerobat's airframe skin is also heftier than the skin on the 150. Cessna hoped to complete its training lineup by offering an aerobatic trainer, but the Aerobat falls short of the mark in certain important respects. The lack of an inverted system and the same airfoil that is used on the 150 greatly limit its abilities as a true aerobatic trainer. However, it is a very good trainer for spins and unusual attitudes, and might also be found quite suitable by the occasional "Sunday looper."

De Havilland Chipmunk

The Chipmunk was designed and developed by De Havilland's Canadian subsidiary in the mid-1940s as a military trainer and was eventually built on both sides of the Atlantic in fairly large numbers. It fulfilled a long military and civilian career worldwide, and is one of the most delightful aircraft to fly. The pilot truly feels as if the Chipmunk's wings are an extension of the human body. Aerobatics are an effortless ballet. The stock airplane's obsolete 140-hp Gypsy Major engine leaves it underpowered for modern, dynamic aerobatics, and the lack of a constant-speed propeller is another drawback. Long out of production, the Chipmunk is a beloved weekend classic, in spite of expensive maintenance and spare parts costs. The many modifications to the airplane by top-level airshow pilots into the 1970s attests to the airframe's highly regarded aerobatic qualities. The most famous of the modified Chipmunks was Art Scholl's Super Chipmunk. If the Chipmunk had been further developed, it is quite likely that a derivative aircraft could have easily shared the limelight with the legendary Czech Zlin Trener.

Walter Extra 230

The Extra 230, built by Extra Flugzeugbau, is the creation of German pilot/builder Walter Extra. The origins of this midwing monoplane can be traced back to the Stephens Akro via Leo Loudenschlager's Laser 200. The 230 first appeared on the aerobatic scene in 1984 and was brought to the United States by Clint McHenry who won the 1986 U.S. nationals in it. The Extra 230 is a superb unlimited machine with featherlight and exquisitely harmonious controls. It has become a popular competition aircraft.

Walter Extra 300

The Extra 300 is Walter Extra's state-of-the-art two-seat unlimited category machine. As of this writing, it is the only aircraft of its caliber that has certification in the United States (competitors are in the experimental category), paving the way for being operated by flight schools commercially. It was the first of the current generation of unlimited aerobatic aircraft in which unrestricted dual aerobatics was possible (a role subsequently shared after introduction of the Sukhoi 29). It is also the first competition aerobatic aircraft to be made almost entirely of composites. The Extra 300 was introduced to North America in 1988 at the World Aerobatic Championships in Red Deer, Canada. Its versatility as a competition machine as well as an advanced instructional aircraft should ensure the Extra 300 a niche in the skies for many years.

Walter Extra 300S

The Extra 300S is Walter Extra's answer to such competitors as the Russian Sukhoi 26, the French CAP 231, and the American Staudacher 300 series. Introduced to the competition scene at the 1992 World Championship in France, the 300S has an astounding roll rate and retains all the outstanding qualities that have made the Extra line so popular. Expect it to be a top competition machine for a long time to come.

Sukhoi 26

The Sukhoi 26 is one of the world's premiere unlimited aerobatic aircraft. It created a big stir when it made its first appearance at the 1984 World Championships in Bekescsaba, Hungary. Still in the prototype stage, it exhibited a phenomenal roll rate and impressed the crowds and officials alike by easily performing five successive vertical rolls. Refinements came in the following years, the appointment of a United States dealer, and a warm welcome at Oshkosh in 1989. Soviet when it went into production, the Sukhoi 26 ended up Russian when the Soviet Union collapsed on itself.

It is produced by the same factory whose main Cold War business was building one of the world's most formidable supersonic fighters. No resources of the company were spared in designing the Sukhoi 26. The mixed construction of its immensely strong airframe includes composite wings and exotic metals in the fuselage. Powered by a 360-hp Vedeneyev M-14 9-cylinder radial engine, it has a stunning roll rate, excellent vertical performance, and is also capable of astounding slow-speed maneuvers. An innovative feature on the Sukhoi 26 is the partially reclining pilot's seat (à la F-16) which makes it noticeably easier to withstand high G forces.

Sukhoi 29

In the eyes of many would-be unlimited aerobatic pilots, the Sukhoi 26 had just one problem. It had only one seat, a scary proposition for the inexperienced. This shortcoming was solved by the two-seat Sukhoi 29. It can do everything the Sukhoi 26 can do, but with two on board. Though sharing some common elements with the Sukhoi 26, the Sukhoi 29 is essentially a completely different aircraft with an all-composite monocoque fuselage and a different wing. It is the aerobatic unlimited two-seater par excellence, and a formidable competition aircraft flown solo.

Staudacher S300

The Staudacher S300 unlimited aerobatic monoplane is custom built by John Staudacher in Saginaw, Michigan. It captivated the aerobatic community with its excellent performance and craftsmanship when it made its public debut in 1990. In 1992, Michael Goulian placed second at the Unlimited U.S. Nationals in the prototype Staudacher S300. Roll rate and vertical performance are equal to the abilities of the top European and Russian unlimited production monoplanes. The S300 has a spruce-and-carbon-fiber wing and a traditional steel-tube-and-fabric fuselage. John Staudacher was one of America's preeminent hydroplane racing boat designers and builders before he turned his talents to designing and building unlimited aerobatic competition aircraft.

Staudacher 300GS

The Staudacher 300GS was developed specifically for Mike Goulian by John Staudacher from the S300. Changes include a raised thrustline, shortened wing, lengthened ailerons, and a more streamlined canopy. According to coauthor Goulian, the machine is arguably the world's finest unlimited aerobatic airplane as of this writing.

Zlin 226/326

The Czech tandem two-seat metal-and-fabric Zlin 226 monoplane electrified the world of aerobatics in the late 1950s when, in the hands of the Czech aerobatic team, it changed the style of competition flying to what has since become known as *dynamic aerobatics*. The sequences performed in the 226 were laden with snap rolls, spins, negative maneuvers, and a lot of inverted flying. The ultimate gyroscopically induced autorotation maneuver, the Lomcovak, was first flown and perfected in a Zlin. The 226 was further refined in the 326 Trener Master with the addition of retractable gear. Paramilitary basic training in preparation for entry into an air force was an important role for the Trener Masters, a result of the standard two-seat configuration; however, for aerobatic competition, single-seat versions (all known as Akrobats) were also developed.

Zlin 526

The Zlin 526 Trener Master was the final version of the Trener line. It was produced in a variety of versions from 1965 into the 1970s. The biggest improvements compared to the 326 were the addition of a constant-speed propeller and a revised wing. The later 526s also had a more powerful, 180-hp engine. Most 526s were two-seaters, but a limited number of single-seaters (also known as Akrobats) were also produced. Most of the Zlin Treners registered in the United States are late-model 526s.

Zlin 50S

The Zlin 50 was the Czech answer to the aircraft types that began to displace the Trener series on the competition scene in the early 1970s. It appeared in 1975 and was a completely new design. Equipped with a 260-hp Lycoming that was later upgraded to 300 hp, and a three-blade constant-speed Hoffman propeller, it had a phenomenal roll rate and could do four vertical rolls. Pilots accustomed to an earlier generation of unlimited machines took some time getting used to the lightning quick ailerons; the rudder is considered to be a bit on the heavy side. Nevertheless, for the next 10 years following its introduction, the Zlin 50 together with the Yak 50 dominated the top placings in international aerobatic competition. It is now being outperformed by a newer generation of monoplanes, but still remains a formidable unlimited aircraft.

Stearman

The Stearman is one of the grand old antiques, but it is somewhat younger than its lines would suggest. It is also really a Boeing. Despite its-1920s looks, it didn't go into production until 1935. By then, Stearman was a subsidiary of Boeing. More than 8,000 Stearmans were built go get a whole generation of pilots on their way to World War II. More than 2,000 Stearmans are still flying today. Originally equipped with a 220-hp engine, the Stearman is a solid, honest airplane in which old fashioned, classic aerobatics are one big, delightful nostalgia trip. It is also a great crowd pleaser and is widely seen on the airshow display circuit. It has a reputation for being as solid as a brick and is often reengined with engines as powerful as 500+ hp.

Stampe SV4

The Stampe was once one of the most popular aerobatic airplanes in Europe, a role it pinched from the Tiger Moth after World War II when the Stampe was bought by the French government in great numbers for its air force and its national network of subsidized aeroclubs. The SV4 design was actually completed in 1937 and selected as a basic trainer by the French Air Force in 1939, but World War II intervened. Its Belgian designer was strongly inspired by the Tiger Moth and the designer's new airplane succeeded in eliminating many of the Tiger's shortcomings. The Stampe's two sets of ailerons make it one of the more agile aerobatic biplanes. It requires precision to be flown well, but is forgiving of mistakes, and is still an excellent teacher of basic, old-fashioned aerobatics. The Stampe was manufactured in limited numbers in Belgium and mass-produced in France and Algeria. Though not imported into the United States in great numbers, it is possible to find one of these pleasant airplanes if one looks hard enough.

Bücker Jungmann

The Bücker Jungmann was introduced in 1934 in Germany and was soon an international best-seller, produced under license in Switzerland and Spain. It had an important training role in the German Luftwaffe through World War II. Although German production did not resume in the postwar years, the Jungmann continued to be built in Spain until 1960 and was revived in Switzerland as the Lerche with a 180-hp Lycoming engine and an improved airfoil as late as 1966. The agile, powerful Jungmann is small for a biplane. It is still an outstanding basic aerobatic trainer. Quite a few Jungmanns found their way to the Unites States; among the many well-known American pilots who cherished their own Jungmann was the aviation author Ernest K. Gann.

Bücker Jungmeister

The Jungmeister, introduced in 1936, was the Bücker Jungmann's single-seat cousin and considered by many to be one of the best aerobatic airplanes of all time. With its small, light airframe, powerful 160-hp engine, and crisp controls, it can be thought of as a forerunner of the Pitts Specials. It took top honors in international competition soon after it was introduced. It has become a sought-after antique, and continues to delight airshow spectators to this day.

De Havilland Tiger Moth

The venerable Tiger is the most successful of Geoffrey De Havilland's long line of Moths. Put into production in the early 1930s, it introduced generation after generation of civilian and military pilots throughout the British Commonwealth to the world of flying. Given this role, it may be rightly regarded as Britain's Stearman. The Tiger quite ably flies all the basic aerobatic maneuvers that were essential elements of the military curriculum of its day, but the truth be said, its draggy airframe, sloppy controls, and low power make it one of the less inspiring aerobatic machines, and it never amounted to much in competition. It is a fantastic and much-beloved nostalgia machine, steadily increasing in value, but if serious antique biplane aerobatics are your interest, get a Stampe, a Bücker, or a Stearman. Many of the Tiger Moths in the United States were imported from India when they were retired from the Indian Air Force in the mid-1970s.

Acrostar

The Acrostar is an intriguing unlimited monoplane developed by Arnold Wagner of Switzerland. It first flew in 1970 and was competitive at the World Championship level for over a decade. The Acrostar has a completely symmetrical wing and a control system that is radical even by today's standards. The *stabilator* (all moving horizontal stabilizer-elevator) is interconnected with the flaps, and the flaps are connected to the ailerons. Perhaps the most accomplished Acrostar pilot was the Swiss flyer Eric Müller, who competed in the perky little machine for 10 years.

Yak 18

There are probably few airplanes that exist in as many versions as the Yak 18. Initially introduced as a basic military trainer in 1946, it has been in service in some form or fashion ever since, and its derivatives are still the main aerobatic trainer of the states that formed the Soviet Union. It has also been in widespread use in many other countries of the former Communist bloc. The popular civilian aerobatic trainer version is a tricycle gear model that has been in use since the early 1960s. Single-seat taildragger variants were developed for unlimited competition and were the mainstay of Soviet aerobatic competitors until replaced by the Yak 50, and more recently the Yak 55.

The Yak 18 is a big, heavy airplane, and most competitive versions were equipped with 300-hp engines back when such horsepower was rarely found on competition aircraft. Since the cessation of Cold War tensions, a handful of Yak 18s have been finding their way to the United States, most notably from China and Hungary. Flown hard, a few Yak 18s have suffered spar failures, one of which claimed the life of one of the best Russian competition pilots.

Yak 50/52

The Yak 50 replaced the Yak 18 as the standard competitive aerobatic aircraft in what was then the Soviet Union. It is a direct descendant of the Yak 18 with an updated wing and a 360-hp Vedeneyev M-14 engine (which also powers the Sukhoi airplanes) among other improvements. It first appeared in 1975 and for the next decade it shared most international top honors with the Czech Zlin 50. Even more powerful than its predecessor, the Yak 50 has outstanding vertical performance even by today's standards, though it is no longer competitive as an unlimited machine.

Yak 55

As the 1970s came to a close, the designers at the Yakovlev Design Bureau realized that they had milked all there was to be had from the Yak 18 and its derivative, the Yak 50. They came up with a completely new airplane, the Yak 55. It was a promising design. Small by Russian standards, lightweight with a big, fat, symmetrical airfoil, it was powered by the 360-hp Vedeneyev engine. It was, however, overshadowed by the Sukhoi 26, which was even smaller and lighter, and outperformed the Yak 55 in roll rate and vertical performance.

Stephens Akro

This graceful little midwing monoplane has a landmark place in aerobatic history because although not built in great quantities, it directly inspired several of the present generation of unlimited monoplanes and is responsible more than any type for the standards of today's unlimited machines. The Akro was developed in 1967 by Clayton Stephens for the late Margaret Ritchie, who perished in the prototype due to causes unrelated to the integrity of the aircraft.

The Stephens Akro was catapulted into the limelight by Leo Loudenschlager when he embarked on his competition career that would eventually lead to seven unlimited national championships and a world championship title. The Stephens Akro was always plan-built, which provided ample opportunity for modification and evolution. Direct derivatives are Loudenschlager's famous Laser 200 (the next airplane description in this chapter) and Henry Haigh's Superstar, both world championship winners, and the Extra 230.

Laser 200

The Laser 200 is Leo Loudenschlager's redesign of his Stephens Akro. Leo had flown the Stephens Akro for four years by 1975 and felt he could greatly improve on it. The result was the Laser 200. It had a new airfoil and a modified, considerably lighter fuselage. The high power loading and roll rate enabled Loudenschlager to outperform the Pitts Specials then dominating the United States competition scene. In 1980, Loudenschlager became world champion in his Laser 200. He is retired from competition aerobatics but the Laser 200, which sports a distinctive bright red Budweiser color scheme, continues to delight the airshow crowds with Loudenschlager at the controls.

Steen Skybolt

The Steen Skybolt is a popular homebuilt biplane, ideal for the serious recreational aerobatic pilot. It is also a fun competition airplane at the Sportsman level. The Skybolt is a traditional wood, welded steel, and fabric airplane and superficially looks like a two-seat Pitts. The design has been around for decades and there are quite a few examples flying.

Slingsby Firefly

The Firefly is the most recent derivative of Slingsby's successful side-by-side two-seat trainer/tourer. It is unsuited for aerobatic competition, but is an outstanding basic trainer. Its maker, Slingsby is best known for a long line of gliders, a heritage that shows in the Firefly's graceful low-drag composite airframe, generous bubble canopy, and high aspect ratio wings. The Firefly is a composite version of its predecessor, the wooden T-67, which has been flying since the mid-1970s. The two airframes look identical. Various engine and propeller combinations are available. The U.S. Air Force selected the Slingsby Firefly as its primary trainer in 1992.

Taperwing Waco

The Taperwing initially started as a marketing idea to jazz up the straight-wing Waco's staid looks, and ended up as a popular aerobatic biplane of its generation. The tapering shape increased the wing's efficiency and rolling performance and, combined with a powerful engine, gave the airplane outstanding vertical performance for its time. Though by today's standards its aerobatic capabilities are unremarkable, in its time it was a favorite airplane of professional display pilots like Freddie Lund, Phil Love, and Tex Rankin. The trio's Taperwing Stunt Team was a much sought-after airshow act. Taperwings are rare, but if you really want to own and fly perhaps the greatest Golden Age antique, you might still be able to find one.

Great Lakes

The two-seat Great Lakes biplane was first produced in 1929 and fell victim to the depression in 1933. It was small for a biplane in those days and excelled at aerobatics. Over the years it was refined, and reengined. An enlargement of the rudder led to a popular distinction between "small tail" and "large tail" Great Lakes. Difficulties with obtaining parts for its Cirrus engines diminished their popularity with flight schools. The Great Lakes was a favorite mount of Tex Rankin the colorful airshow pilot who flew 131 outside loops in as many minutes in the airplane. Today the Great Lakes is a valued antique, though often reengined with a Lycoming or a Warner Scarab. The Great Lakes has been produced twice since the early seventies, albeit briefly each time.

Siai Marchetti SF .260

The stylish Italian Siai Marchetti SF .260 has been in production since 1964. It is widely used as a military trainer as well as a civilian hot-rod. It is fully aerobatic, though not a competition airplane. Civilian owners can have great fun doing military-style flying in the SF .260, and it is a fast, but expensive, cross-country machine. SF .260s are also popular with airshow display pilots. The SF .260 was at one point marketed in the United States as the Waco Meteor.

Beech T-34

The Beech T-34 is another airplane in which civilians have the opportunity to enjoy the type of flying done in military flight training, including aerobatics (though the airplane is a bit on the heavy side for anything but the basics). The U.S. Air Force and U.S. Navy have operated T-34s as trainers; the Navy is flying a turboprop version. Many civilians feel quite at home in the T-34 after an hour or so because the airplane is really nothing more than a beefed up Bonanza with a tandem two-seat cockpit. It is tremendous fun as a civilian recreational airplane, and can put on an outstanding airshow act, but is not a good choice if your interest is competitive aerobatics.

IAC Onedesign

The International Aerobatic Club is developing this affordable kit-built competition machine. Contact IAC for more information about the Onedesign, which is a low-wing monoplane.

Beech F-33C Bonanza

Approximately 50 aerobatic Bonanzas were built as trainers for foreign air forces. About a dozen of the four-seat aircraft are flying in the United States.

22
Aerobatic organizations

AEROBATICS HAS COME A LONG WAY FROM ITS DAREDEVIL BEGINNINGS
to being a safe, highly organized sport. Today it is supported by a network of organizations worldwide that provide standards, guidance, and an operating framework for everyone from the grass roots level recreational aerobatic pilot to the international level unlimited aerobatic competitor.

At the international level, aerobatics is overseen by CIVA, the Federation Internationale Aeronautique's aerobatic organization. CIVA, a body of official representatives of its constituent countries, oversees the world championships. It establishes and supervises the rules by which world competitions are held and judged. CIVA's international competition rules are not binding on member nations at the national level but serve as a voluntary model for the national aerobatic organizations around the world, to be adapted, implemented, and modified as each organization sees fit.

The United States' national organization supporting aerobatics is the International Aerobatic Club (IAC), which is a division of the Experimental Aircraft Association (EAA). Membership is open to anyone. IAC membership in 1993 was over 5,000. The National Aeronautic Association, which is the United States' liaison organization to the FAI, officially appointed the IAC in 1991 to administer all aerobatic activity in the United States.

The IAC has responsibilities in three areas:

- Promotion and support of grass roots aerobatics through a nationwide network of local IAC chapters.

- Administration of national competitions and the sanctioning of regional and local competitions.

- Organization and management of the U.S. team's participation in the world championships; fund raising for the U.S. team and administration of U.S. Aerobatic Foundation.

Fig. 22-1. *Expect to find like-minded pilots with airplanes such as yours at the local IAC chapter.*

The IAC provides many services directly, plus it supports the activities of its local chapters. Specifically, the IAC:

- Oversees an Achievement Awards Program administered by the local chapters. This is a grass roots level program designed to provide recognition to aerobatic pilots as they accomplish various levels of competence.
- Publishes *Sport Aerobatics*, an excellent periodical magazine that is the main forum of communication for the U.S. aerobatic community.

- From time to time publishes volumes of technical tips, compiling useful safety and technical operating material for all phases of the sport and covering many aerobatic aircraft.
- Establishes the rules for competition at all levels; makes rule books and CIVA regulations available to all members.
- Sets the annual known sequences for all categories of competition, except the unlimited known category, which is set by CIVA.
- Organizes competitions at the national level, culminating in the U.S. National Championships in the fall at the end of the competition season.
- Sanctions and oversees approximately 40–50 regional and local competitions during each competition season.
- Administers a judges' continuing education program and a judges' home study program.
- Supports aerobatic safety seminars nationwide.
- Maintains a nationwide list of aerobatic schools available to anyone (The IAC does not endorse any of the schools. The fact that a school is on IAC's list is no indication of any quality judgment by IAC).
- Publishes a very useful annual membership directory.

At the heart of grass roots aerobatics in the United States is the nationwide network of approximately 50 local IAC chapters. They operate along the line of franchises. They are given organizational assistance and continuing support by the IAC, but are essentially independent local entities. The address of the local chapter is available to anyone from the IAC.

Each IAC chapter organizes regular meetings where aerobatic matters of all kinds are discussed, publishes a periodic newsletter, and organizes aerobatic technical and safety seminars and judges' schools in conjunction with IAC. IAC chapters also organize the local and regional competitions.

Above all, the local IAC chapter is a place to turn to for the aerobatic pilot seeking camaraderie and support to make the most of one of the most demanding and rewarding sports in the world.

Appendix A

Ten-hour aerobatic course

Each lesson is identified by the new maneuvers introduced during the lesson; however, an important component of each lesson is the continued practice and refinement of maneuvers introduced during previous lessons, as outlined in the lesson plans. Each lesson is conducted on a suggested time schedule:

- Preflight training: 45 minutes
- Flight session: Demonstration 15 minutes
- Student practice: 30 minutes
- Postflight debriefing: 15 minutes
- Total time: 2 hours

LESSON 1: STALLS, AILERON ROLL

Objective

Introduction to tailwheel aircraft, familiarization with aerobatic preflight, confidence building flight exercises, stall theory, performance of stalls (power on, power off, accelerated, banked). Introduction of the aileron roll as the first aerobatic maneuver (unusual attitude experience in low, positive G conditions).

Preflight training

Introduce aerobatic preflight, discuss aerodynamics and limits as they apply to aerobatic flight; discuss stalls; definition, causes, aircraft design characteristics, technique. Discuss aileron roll theory and technique.

Flight session

1. Rudder coordination exercises
2. Turns, MCA, turns at MCA
3. Stalls (all configurations)
4. Aileron rolls

Completion standards

Ability to perform stalls; recover confidently from all attitudes. Ability to perform aileron rolls with minor errors.

LESSON 2: SLOW ROLLS, INVERTED FLIGHT

Objective

Introduction of the sensation of inverted flight and negative G through the slow roll and the half slow roll to inverted. Continued practice of stalls.

Preflight training

Discuss slow roll theory and technique, the aerodynamics of inverted flight and inverted flight technique, including turns when inverted.

Flight session

1. MCA (review)
2. Stalls (review)
3. Slow rolls
4. Half rolls to and from inverted
5. Inverted straight and level

Completion standards

Confident performance of stalls; ability to at least manhandle the aircraft around in a slow roll attempt unassisted; ability to recover from an inverted attitude.

If the lesson is completed with minimal effort, the loop may be introduced on the way home.

LESSON 3: LOOPS, HALF LOOPS

Objective

Introduction of the loop; half loop with a transition to prolonged inverted flight followed by a roll to upright. Continued practice and refinement of slow rolls and half rolls; shallow turns inverted.

Preflight training

Discuss theory and technique of loops and half loops. Review slow roll and half roll errors committed during previous session and corrective action required.

Flight session

1. Slow rolls (review)
2. Half rolls (review)
3. Loops
4. Half loops (extended inverted, straight and level, shallow turns)

Completion standards

Ability to safely perform loops with minimum altitude loss and heading correction, and half loops with smooth transition to inverted straight and level with good attitude control and minimum heading correction; ability to safely perform slow rolls and half rolls with medium precision; ability to recognize slow roll errors independently.

LESSON 4: IMMELMANN, HALF CUBAN EIGHT

Objective

Develop the half loop into the Immelmann; introduce the half Cuban eight; continue to perfect the slow roll.

Preflight training

Discuss the Immelmann and half Cuban eight theory and technique with special emphasis on the vital importance of the correct visual references in both maneuvers. Review slow roll and loop errors and corrective action.

Flight session

1. Slow rolls (review)
2. Half rolls (review)
3. Immelmanns
4. Half Cuban eights

Completion standards

Ability to safely and accurately perform Immelmanns; ability to maintain orientation throughout half Cuban eight and safely fly it with medium precision; ability to fly slow rolls safely and precisely.

LESSONS 5 AND 6: REVIEW AND REFINEMENT OF MANEUVERS

Objective

Work on refining learned maneuvers to competition standards; begin to fly maneuvers in sequence.

Preflight training

Review errors and corrective action for maneuvers learned; explain competition standards; introduce maneuver sequencing concepts and technique.

Flight session

1. Slow rolls
2. Half rolls
3. Immelmanns
4. Half Cuban eights
5. Loop, half Cuban, Immelmann sequence

Depending on what maneuvers of the student require further attention, the session may be adjusted to suit individual needs.

Completion standards

All maneuvers should be flown safely and with a precision expected of a beginner Sportsman competitor.

LESSON 7: RECOVERY FROM INCOMPLETE MANEUVERS. REVERSE HALF CUBAN EIGHT

Objective

Develop ability to recover from incomplete maneuvers—maneuvers that have gone wrong; introduce reverse half Cuban eight.

Preflight training

Discuss how maneuvers go wrong, how to recognize in time that a maneuver is about to go wrong, and what corrective action to take (theory and technique); discuss reverse half Cuban eight theory and technique.

Flight session

1. Slow rolls (review)
2. Half rolls (review)

3. Inverted straight and level (review)
4. Reverse half Cuban eight
5. Emergency recoveries

Completion standards

Ability to perform all maneuvers safely and precisely; ability to recover with confidence from maneuvers gone wrong.

LESSON 8: SPINS

Objective

Develop the ability to perform intentional spins, competition spins; learn emergency spin recovery. Continue to refine other maneuvers.

Preflight training

Detailed discussion of spin theory and technique; special emphasis on emergency spin recovery.

Flight session

1. Slow rolls (review)
2. Half rolls (review)
3. Half Cubans (review)
4. Reverse half Cubans (review)
5. Spins

Select review maneuvers based on individual student need. Allow sufficient flight time for thorough spin instruction.

Completion standards

Ability to enter and recover from spins safely on approximate heading; ability to safely recover from inadvertent spins; ability to perform all other maneuvers to beginning Sportsman standard.

LESSON 9: HAMMERHEADS

Objective

Learn to perform the hammerhead; continue to refine the other maneuvers to Sportsman level competition standards.

Preflight training

Discuss hammerhead theory and technique; review errors in performing other maneuvers and discuss corrective action required.

Flight session

1. Slow rolls (review)
2. Half rolls (review)
3. Reverse half Cubans (review)
4. Spins (review)
5. Hammerheads

Completion standards

Ability to perform hammerheads safely; ability to recover from spins precisely on heading; ability to perform all other maneuvers to Sportsman standards.

LESSON 10: FINAL REVIEW

Objective

Review all maneuvers learned and string together a typical sportsman style sequence.

Preflight training

Discuss errors and corrective action required; lay out sequence and discuss flying the sequence.

Flight session

Sample sequence: Loop, half Cuban eight, roll, hammerhead, reverse half Cuban eight, Immelmann, spin.

Completion standards

Ability to safely practice all maneuvers solo and perform them to Sportsman competition standards.

Appendix B

Five-hour aerobatic course

Each lesson is identified by the new maneuvers introduced during the lesson; however, an important component of each lesson is the continued practice and refinement of maneuvers introduced during previous lessons, as outlined in the lesson plans. Each lesson is conducted on a suggested time schedule:

- Preflight training: 45 minutes
- Flight session: Demonstration 15 minutes
- Student practice: 30 minutes
- Postflight debriefing: 15 minutes
- Total time: 2 hours

LESSON 1: STALLS, AILERON ROLL

Objective

Introduction to tailwheel aircraft, familiarization with aerobatic preflight, confidence building flight exercises, stall theory, performance of stalls (power on, power off, accelerated, banked). Introduction of the aileron roll as the first aerobatic maneuver (unusual attitude experience in low, positive G conditions).

Preflight training

Introduce aerobatic preflight, discuss aerodynamics and limits as they apply to aerobatic flight; discuss stalls; definition, causes, aircraft design characteristics, technique. Discuss aileron roll theory and technique.

Flight session

1. Rudder coordination exercises
2. Turns, MCA, turns at MCA
3. Stalls (all configurations)
4. Aileron rolls

Completion standards

Ability to perform stalls; recover confidently from all attitudes. Ability to perform aileron rolls with minor errors.

LESSON 2: SLOW ROLLS, INVERTED FLIGHT

Objective

Introduction of the sensation of inverted flight and negative G through the slow roll and the half slow roll to inverted. Continued practice of stalls.

Preflight training

Discuss slow roll theory and technique, the aerodynamics of inverted flight and inverted flight technique, including turns when inverted.

Flight session

1. MCA (review)
2. Stalls (review)
3. Slow rolls
4. Half rolls to and from inverted
5. Inverted straight and level

Completion standards

Confident performance of stalls; ability to at least manhandle the aircraft around in a slow roll attempt unassisted; ability to recover from an inverted attitude.

If the lesson is completed with minimal effort, the loop may be introduced on the way home.

LESSON 3: LOOPS, HALF LOOPS

Objective

Introduction of the loop; half loop with a transition to prolonged inverted flight followed by a roll to upright. Continued practice and refinement of slow rolls and half rolls; shallow turns inverted.

Preflight training

Discuss theory and technique of loops and half loops. Review slow roll and half roll errors committed during previous session and corrective action required.

Flight session

1. Slow rolls (review)
2. Half rolls (review)
3. Loops
4. Half loops (extended inverted, straight and level, shallow turns)

Completion standards

Ability to safely perform loops with minimum altitude loss and heading correction, and half loops with smooth transition to inverted straight and level with good attitude control and minimum heading correction; ability to safely perform slow rolls and half rolls with medium precision; ability to recognize slow roll errors independently.

LESSON 4: IMMELMANN, HALF CUBAN EIGHT

Objective

Develop the half loop into the Immelmann; introduce the half Cuban eight; continue to perfect the slow roll.

Preflight training

Discuss the Immelmann and half Cuban eight theory and technique with special emphasis on the vital importance of the correct visual references in both maneuvers. Review slow roll and loop errors and corrective action.

Flight session

1. Slow rolls (review)
2. Half rolls (review)
3. Immelmanns
4. Half Cuban eights

Completion standards

Ability to safely and accurately perform Immelmanns; ability to maintain orientation throughout half Cuban eight and safely fly it with medium precision; ability to fly slow rolls safely and precisely.

LESSON 5: SPINS

Objective

Develop the ability to perform intentional spins, competition spins; learn emergency spin recovery. Continue to refine other maneuvers.

Preflight training

Detailed discussion of spin theory and technique; special emphasis on emergency spin recovery.

Flight session

1. Slow rolls (review)
2. Half rolls (review)
3. Immelmanns (review)
4. Half Cubans (review)
5. Spins

Completion standards

Ability to enter and recover from spins safely on approximate heading; ability to safely recover from inadvertent spins; ability to safely practice all maneuvers solo.

Appendix C
Organizations and aircraft manufacturers

International Aerobatic Club and
Experimental Aircraft Association
P.O. Box 3086
Oshkosh, Wisconsin 54903-3086
414-426-4800

AVIAT, Inc. (Pitts, Christen aircraft)
P.O. Box 1149
Afton, Wyoming 83110
307-886-3151

American Champion Aircraft (Decathlon aircraft)
P.O. Box 37
Rochester, Wisconsin 53167
800-223-9381

Advanced Sukhoi Technologies
23A Policarpov St.
Moscow 125284
Russia

U.S. dealer:
Pompano Air Center
1401 Northeast 10th St.
Pompano Beach, Florida 33060
305-943-6050

Staudacher Aircraft
2648 East Beaver Rd.
Kawkawlin, Michigan 48631
517-684-7230

Extra Flugzeugbau
Flugplatz Dinslaken
4224 Hunxe
Germany
011-49-2858-7124

U.S. dealer:
Aero Sport
P.O. Drawer 1989
St. Augustine, Florida 32085
904-842-6230

Avions Mudry et Cie
BP 214 Aerodrome
27300 Bernay
France
33-32-43-47-34

U.S. dealer:
Mudry Aviation, Ltd
Sr 1, Box 18T, #7
Bunnell, Florida 32110
904-437-9700

Slingsby Aviation, Ltd.
Kirkbymoorside,
York YO6 6EZ
England
(0751) 32474

Appendix D

The cost of aerobatic aircraft ownership

A spreadsheet is one way to determine the feasibility of owning an aerobatic aircraft. (This spreadsheet's template is part of *Fly for Less, Flying Clubs and Aircraft Partnerships* (TAB-McGraw Hill, 1992). For additional details on how to analyze and cut the costs of your flying, refer to *Fly for Less*.)

AIRCRAFT: Aerobatic Trainer

AIRCRAFT PARTNERSHIP FINANCIAL ANALYSIS - OWNERSHIP COSTS, 1-15 PILOTS

NUMBER OF PILOTS	1	2	3	4	5	6	7	8	9	10	11	12	13	14	15
ANNUAL FIXED EXPENSES															
Tiedown/Hangar	3000.00	1500.00	1000.00	750.00	600.00	500.00	428.57	375.00	333.33	300.00	272.73	250.00	230.77	214.29	200.00
Insurance	1800.00	900.00	600.00	450.00	360.00	300.00	257.14	225.00	200.00	180.00	163.64	150.00	138.46	128.57	120.00
State Fees	150.00	75.00	50.00	37.50	30.00	25.00	21.43	18.75	16.67	15.00	13.64	12.50	11.54	10.71	10.00
Annual	500.00	250.00	166.67	125.00	100.00	83.33	71.43	62.50	55.56	50.00	45.45	41.67	38.46	35.71	33.33
Maintenance	1000.00	500.00	333.33	250.00	200.00	166.67	142.86	125.00	111.11	100.00	90.91	83.33	76.92	71.43	66.67
Loan Payments	3568.07	1784.03	1189.36	892.02	713.61	594.68	509.72	446.01	396.45	356.81	324.37	297.34	274.47	254.86	237.87
Cost of Capital (non-cash)	1575.00	787.50	525.00	393.75	315.00	262.50	225.00	196.88	175.00	157.50	143.18	131.25	121.15	112.50	105.00
Total Fixed Expenses / yr	11593.07	5796.53	3864.36	2898.27	2318.61	1932.18	1656.15	1449.13	1288.12	1159.31	1053.92	966.09	891.77	828.08	772.87
HOURLY OPERATING EXPENSES															
Fuel	21.00	21.00	21.00	21.00	21.00	21.00	21.00	21.00	21.00	21.00	21.00	21.00	21.00	21.00	21.00
Oil	0.80	0.80	0.80	0.80	0.80	0.80	0.80	0.80	0.80	0.80	0.80	0.80	0.80	0.80	0.80
Engine Reserve	8.00	8.00	8.00	8.00	8.00	8.00	8.00	8.00	8.00	8.00	8.00	8.00	8.00	8.00	8.00
General Maint Res	10.00	10.00	10.00	10.00	10.00	10.00	10.00	10.00	10.00	10.00	10.00	10.00	10.00	10.00	10.00
Total Op Exp / hr	39.80	39.80	39.80	39.80	39.80	39.80	39.80	39.80	39.80	39.80	39.80	39.80	39.80	39.80	39.80
TOTAL HOURLY EXPENSES															
50 Hours	271.66	155.73	117.09	97.77	86.17	78.44	72.92	68.78	65.56	62.99	60.88	59.12	57.64	56.36	55.26
100 Hours	155.73	97.77	78.44	68.78	62.99	59.12	56.36	54.29	52.68	51.39	50.34	49.46	48.72	48.08	47.53
150 Hours	117.09	78.44	65.56	59.12	55.26	52.68	50.84	49.46	48.39	47.53	46.83	46.24	45.75	45.32	44.95
Hourly Commercial Rental	100.00	100.00	100.00	100.00	100.00	100.00	100.00	100.00	100.00	100.00	100.00	100.00	100.00	100.00	100.00
TOTAL ANNUAL EXPENSES															
50 Hours, Own	13583.07	7786.53	5854.36	4888.27	4308.61	3922.18	3646.15	3439.13	3278.12	3149.31	3043.92	2956.09	2881.77	2818.08	2762.87
Own (cash only)	12008.07	6999.03	5329.36	4494.52	3993.61	3659.68	3421.15	3242.26	3103.12	2991.81	2900.73	2824.84	2760.62	2705.58	2657.87
Rent	5000.00	5000.00	5000.00	5000.00	5000.00	5000.00	5000.00	5000.00	5000.00	5000.00	5000.00	5000.00	5000.00	5000.00	5000.00
100 Hours, Own	15573.07	9776.53	7844.36	6878.27	6298.61	5912.18	5636.15	5429.13	5268.12	5139.31	5033.92	4946.09	4871.77	4808.08	4752.87
Own (cash only)	13998.07	8989.03	7319.36	6484.52	5983.61	5649.68	5411.15	5232.26	5093.12	4981.81	4890.73	4814.84	4750.62	4695.58	4647.87
Rent	10000.00	10000.00	10000.00	10000.00	10000.00	10000.00	10000.00	10000.00	10000.00	10000.00	10000.00	10000.00	10000.00	10000.00	10000.00
150 Hours, Own	17563.07	11766.53	9834.36	8868.27	8288.61	7902.18	7626.15	7419.13	7258.12	7129.31	7023.92	6936.09	6861.77	6798.08	6742.87
Own (cash only)	15988.07	10979.03	9309.36	8474.52	7973.61	7639.68	7401.15	7222.26	7083.12	6971.81	6880.73	6804.84	6740.62	6685.58	6637.87
Rent	15000.00	15000.00	15000.00	15000.00	15000.00	15000.00	15000.00	15000.00	15000.00	15000.00	15000.00	15000.00	15000.00	15000.00	15000.00

ASSUMPTIONS:

Hangar/Month:	250.00	Insurance/yr:	1800.00
Fuel Cons (gal/hr):	10.00	Fuel Cost ($/gal):	2.10
Aircraft Value:	45000.00	Loan O/S:	22500.00
Engine MOH Cost:	12000.00	Gen Maint Res/hr:	10.00
State Fees/yr:	150.00	Annual:	500.00
Oil Cons (qt/hr):	0.25	Oil Cost ($/qt):	3.20
Loan Interest Rate:	10.00%	Loan/Inv Term (yrs):	10
Com. Rental/hr:	100.00		
Maintenance/yr:	1000.00		
Time Before OH:	1500.00		
Cost of Capital Rate:	7.00%		

247

AIRCRAFT: Intermediate

AIRCRAFT PARTNERSHIP FINANCIAL ANALYSIS - OWNERSHIP COSTS, 1-15 PILOTS

NUMBER OF PILOTS	1	2	3	4	5	6	7	8	9	10	11	12	13	14	15
ANNUAL FIXED EXPENSES															
Tiedown/Hangar	3000.00	1500.00	1000.00	750.00	600.00	500.00	428.57	375.00	333.33	300.00	272.73	250.00	230.77	214.29	200.00
Insurance	2800.00	1400.00	933.33	700.00	560.00	466.67	400.00	350.00	311.11	280.00	254.55	233.33	215.38	200.00	186.67
State Fees	150.00	75.00	50.00	37.50	30.00	25.00	21.43	18.75	16.67	15.00	13.64	12.50	11.54	10.71	10.00
Annual	500.00	250.00	166.67	125.00	100.00	83.33	71.43	62.50	55.56	50.00	45.45	41.67	38.46	35.71	33.33
Maintenance	1000.00	500.00	333.33	250.00	200.00	166.67	142.86	125.00	111.11	100.00	90.91	83.33	76.92	71.43	66.67
Loan Payments	6739.69	3369.84	2246.56	1684.92	1347.94	1123.28	962.81	842.46	748.85	673.97	612.70	561.64	518.44	481.41	449.31
Cost of Capital (non-cash)	2975.00	1487.50	991.67	743.75	595.00	495.83	425.00	371.88	330.56	297.50	270.45	247.92	228.85	212.50	198.33
Total Fixed Expenses / yr	17164.69	8582.34	5721.56	4291.17	3432.94	2860.78	2452.10	2145.59	1907.19	1716.47	1560.43	1430.39	1320.36	1226.05	1144.31
HOURLY OPERATING EXPENSES															
Fuel	42.00	42.00	42.00	42.00	42.00	42.00	42.00	42.00	42.00	42.00	42.00	42.00	42.00	42.00	42.00
Oil	1.60	1.60	1.60	1.60	1.60	1.60	1.60	1.60	1.60	1.60	1.60	1.60	1.60	1.60	1.60
Engine Reserve	12.00	12.00	12.00	12.00	12.00	12.00	12.00	12.00	12.00	12.00	12.00	12.00	12.00	12.00	12.00
General Maint Res	10.00	10.00	10.00	10.00	10.00	10.00	10.00	10.00	10.00	10.00	10.00	10.00	10.00	10.00	10.00
Total Op Exp / hr	65.60	65.60	65.60	65.60	65.60	65.60	65.60	65.60	65.60	65.60	65.60	65.60	65.60	65.60	65.60
TOTAL HOURLY EXPENSES															
50 Hours	408.89	237.25	180.03	151.42	134.26	122.82	114.64	108.51	103.74	99.93	96.81	94.21	92.01	90.12	88.49
100 Hours	237.25	151.42	122.82	108.51	99.93	94.21	90.12	87.06	84.67	82.76	81.20	79.90	78.80	77.86	77.04
150 Hours	180.03	122.82	103.74	94.21	88.49	84.67	81.95	79.90	78.31	77.04	76.00	75.14	74.40	73.77	73.23
Hourly Commercial Rental	195.00	195.00	195.00	195.00	195.00	195.00	195.00	195.00	195.00	195.00	195.00	195.00	195.00	195.00	195.00
TOTAL ANNUAL EXPENSES															
50 Hours, Own	20444.69	11862.34	9001.56	7571.17	6712.94	6140.78	5732.10	5425.59	5187.19	4996.47	4840.43	4710.39	4600.36	4506.05	4424.31
Own (cash only)	17469.69	10374.84	8009.90	6827.42	6117.94	5644.95	5307.10	5053.71	4856.63	4698.97	4569.97	4462.47	4371.51	4293.55	4225.98
Rent	9750.00	9750.00	9750.00	9750.00	9750.00	9750.00	9750.00	9750.00	9750.00	9750.00	9750.00	9750.00	9750.00	9750.00	9750.00
100 Hours, Own	23724.69	15142.34	12281.56	10851.17	9992.94	9420.78	9012.10	8705.59	8467.19	8276.47	8120.43	7990.39	7880.36	7786.05	7704.31
Own (cash only)	20749.69	13654.84	11289.90	10107.42	9397.94	8924.95	8587.10	8333.71	8136.63	7978.97	7849.97	7742.47	7651.51	7573.55	7505.98
Rent	19500.00	19500.00	19500.00	19500.00	19500.00	19500.00	19500.00	19500.00	19500.00	19500.00	19500.00	19500.00	19500.00	19500.00	19500.00
150 Hours, Own	27004.69	18422.34	15561.56	14131.17	13272.94	12700.78	12292.10	11985.59	11747.19	11556.47	11400.43	11270.39	11160.36	11066.05	10984.31
Own (cash only)	24029.69	16934.84	14569.90	13387.42	12677.94	12204.95	11867.10	11613.71	11416.63	11258.97	11129.97	11022.47	10931.51	10853.55	10785.98
Rent	29250.00	29250.00	29250.00	29250.00	29250.00	29250.00	29250.00	29250.00	29250.00	29250.00	29250.00	29250.00	29250.00	29250.00	29250.00

ASSUMPTIONS:

Hangar/Month:	250.00	Insurance/yr:	2800.00
Fuel Cons (gal/hr):	20.00	Fuel Cost ($/gal):	2.10
Aircraft Value:	85000.00	Loan O/S:	42500.00
Engine MOH Cost:	18000.00	Gen Maint Res/hr:	10.00

State Fees/yr:	150.00	Maintenance/yr:	1000.00
Oil Cons (qt/hr):	0.50	Time Before OH:	1500.00
Loan Interest Rate:	10.00%	Cost of Capital Rate:	7.00%
Com. Rental/hr:	195.00		

Annual:	500.00		
Oil Cost ($/qt):	3.20		
Loan/Inv Term (yrs):	10		

AIRCRAFT: Unlimited

AIRCRAFT PARTNERSHIP FINANCIAL ANALYSIS - OWNERSHIP COSTS, 1-15 PILOTS

NUMBER OF PILOTS	1	2	3	4	5	6	7	8	9	10	11	12	13	14	15
ANNUAL FIXED EXPENSES															
Tiedown/Hangar	3000.00	1500.00	1000.00	750.00	600.00	500.00	428.57	375.00	333.33	300.00	272.73	250.00	230.77	214.29	200.00
Insurance	4000.00	2000.00	1333.33	1000.00	800.00	666.67	571.43	500.00	444.44	400.00	363.64	333.33	307.69	285.71	266.67
State Fees	150.00	75.00	50.00	37.50	30.00	25.00	21.43	18.75	16.67	15.00	13.64	12.50	11.54	10.71	10.00
Annual	800.00	400.00	266.67	200.00	160.00	133.33	114.29	100.00	88.89	80.00	72.73	66.67	61.54	57.14	53.33
Maintenance	1200.00	600.00	400.00	300.00	240.00	200.00	171.43	150.00	133.33	120.00	109.09	100.00	92.31	85.71	80.00
Loan Payments	11893.57	5946.78	3964.52	2973.39	2378.71	1982.26	1699.08	1486.70	1321.51	1189.36	1081.23	991.13	914.89	849.54	792.90
Cost of Capital (non-cash)	5250.00	2625.00	1750.00	1312.50	1050.00	875.00	750.00	656.25	583.33	525.00	477.27	437.50	403.85	375.00	350.00
Total Fixed Expenses / yr	26293.57	13146.78	8764.52	6573.39	5258.71	4382.26	3756.22	3286.70	2921.51	2629.36	2390.32	2191.13	2022.58	1878.11	1752.90
HOURLY OPERATING EXPENSES															
Fuel	52.50	52.50	52.50	52.50	52.50	52.50	52.50	52.50	52.50	52.50	52.50	52.50	52.50	52.50	52.50
Oil	6.40	6.40	6.40	6.40	6.40	6.40	6.40	6.40	6.40	6.40	6.40	6.40	6.40	6.40	6.40
Engine Reserve	16.67	16.67	16.67	16.67	16.67	16.67	16.67	16.67	16.67	16.67	16.67	16.67	16.67	16.67	16.67
General Maint Res	15.00	15.00	15.00	15.00	15.00	15.00	15.00	15.00	15.00	15.00	15.00	15.00	15.00	15.00	15.00
Total Op Exp / hr	90.57	90.57	90.57	90.57	90.57	90.57	90.57	90.57	90.57	90.57	90.57	90.57	90.57	90.57	90.57
TOTAL HOURLY EXPENSES															
50 Hours	616.44	353.50	265.86	222.03	195.74	178.21	165.69	156.30	149.00	143.15	138.37	134.39	131.02	128.13	125.62
100 Hours	353.50	222.03	178.21	156.30	143.15	134.39	128.13	123.43	119.78	116.86	114.47	112.48	110.79	109.35	108.10
150 Hours	265.86	178.21	149.00	134.39	125.62	119.78	115.61	112.48	110.04	108.10	106.50	105.17	104.05	103.09	102.25
Hourly Commercial Rental	0.00	0.00	0.00	0.00	0.00	0.00	0.00	0.00	0.00	0.00	0.00	0.00	0.00	0.00	0.00
TOTAL ANNUAL EXPENSES															
50 Hours, Own	30821.90	17675.12	13292.86	11101.72	9787.05	8910.59	8284.56	7815.03	7449.84	7157.69	6918.66	6719.46	6550.92	6406.45	6281.24
Own (cash only)	25571.90	15050.12	11542.86	9789.22	8737.05	8035.59	7534.56	7158.78	6866.51	6632.69	6441.38	6281.96	6147.07	6031.45	5931.24
Rent	0.00	0.00	0.00	0.00	0.00	0.00	0.00	0.00	0.00	0.00	0.00	0.00	0.00	0.00	0.00
100 Hours, Own	35350.23	22203.45	17821.19	15630.06	14315.38	13438.93	12812.89	12343.36	11978.17	11686.02	11446.99	11247.80	11079.25	10934.78	10809.57
Own (cash only)	30100.23	19578.45	16071.19	14317.56	13265.38	12563.93	12062.89	11687.11	11394.84	11161.02	10969.72	10810.30	10675.40	10559.78	10459.57
Rent	0.00	0.00	0.00	0.00	0.00	0.00	0.00	0.00	0.00	0.00	0.00	0.00	0.00	0.00	0.00
150 Hours, Own	39878.57	26731.78	22349.52	20158.39	18843.71	17967.26	17341.22	16871.70	16506.51	16214.36	15975.32	15776.13	15607.58	15463.11	15337.90
Own (cash only)	34628.57	24106.78	20599.52	18845.89	17793.71	17092.26	16591.22	16215.45	15923.17	15689.36	15498.05	15338.63	15203.74	15088.11	14987.90
Rent	0.00	0.00	0.00	0.00	0.00	0.00	0.00	0.00	0.00	0.00	0.00	0.00	0.00	0.00	0.00

ASSUMPTIONS:

Hangar/Month: 250.00	Insurance/yr: 4000.00	State Fees/yr: 150.00	Annual: 800.00	Maintenance/yr: 1200.00
Fuel Cons (gal/hr): 25.00	Fuel Cost ($/gal): 2.10	Oil Cons (qt/hr): 2.00	Oil Cost ($/qt): 3.20	Time Before OH: 1200.00
Aircraft Value: 150000.00	Loan O/S: 75000.00	Loan Interest Rate: 10.00%	Loan/Inv Term (yrs): 10	Cost of Capital Rate: 7.00%
Engine MOH Cost: 20000.00	Gen Maint Res/hr: 15.00	Com. Rental/hr: 0.00		

Index